# Mathematics and Culture VI

Michele Emmer (Ed.)

# Mathematics and Culture VI

 Springer

*Editor*
Michele Emmer
Dipartimento di Matematica "G. Castelnuovo"
Università degli Studi "La Sapienza", Roma, Italy
Piazzale Aldo Moro 2
00185 Roma, Italy
email: emmer@mat.uniroma1.it

Translation of the Italian language edition "Matematica e cultura 2006" edited by Michele Emmer, Copyright © Springer-Verlag Italia, Milano 2006.

ISBN  978-3-540-87568-0                    e-ISBN  978-3-540-87569-7

DOI  10.1007/978-3-540-87569-7

Library of Congress Number: 2008943007

Mathematics Subject Classification (2000): 00AXX, 00B10, 01XX, 97AXX

© 2009 Springer-Verlag Berlin Heidelberg

*Typesetting and production*: le-tex publishing services oHG, Leipzig, Germany
*Cover design*: WMXDesign GmbH, Heidelberg, Germany

Engraving on cover and part beginnings by Matteo Emmer, from the book: M. Emmer "La Venezia perfetta", Centro Internazionale della Grafica, Venezia, 1993; by kind permission.

Printed on acid-free paper

9 8 7 6 5 4 3 2 1

springer.com

# Introduction

## Dreaming

Some years ago, in a corridor, in a small space, off to the side, as in a hidden ravine, they were there. They couldn't be anywhere else. Hidden and mysterious, with faces that were dreamy and abstracted, or distracted, or pensive. Caught up in their thoughts, caught up in their space, a space that was distant and that only they could understand. Elusive and yet there, in front of me. Certainly, it was them, the six mathematicians of the *Mathematica* series by Mimmo Paladino. Thinkers of numbers and shapes. For ten years we have been searching for the mathematicians there. In Venice, the favourite place.

An aura of mystery surrounds them, otherwise what kind of mathematicians would they be!

Mysterious, dreamy, absent, absorbed are the faces of Paladino's mathematicians.

The voices of five Sardinian shepherds intone the *Kyrie*, the *Libera Me Domine*, the *Sanctus*. Insistent, profound, archaic voices. The soundtrack of a journey, of a journey towards nothingness. A journey towards *The Wild Blue Yonder*, the last film by Werner Herzog, winner of the international Critic's prize at the Venice festival in 2005. The heroes were astronauts, even if by now no one cares about their adventures; astronauts who take off, who travel, but don't know where they are going.

And then there are the true heroes, the real gurus of the film, the characters who some years ago burst onto the scene in cinema: the mathematicians.

Here they are, the *real* mathematicians of NASA, who do what mathematicians do: they write equations on the blackboard and explain how, by using the gravity of the planets to increase velocity, it would be possible to exit the solar system in order to penetrate into deep space. It is no longer actors who impersonate mathematicians, but the mathematicians themselves who are the heroes. And they ask an artist to *make visible* their scientific dream.

Herzog, in the credits at the end of the film, thanks the astronauts, thanks NASA, and thanks the mathematicians, for their *sense of poetry*.

The endless (which is different from the infinite) journey into the space of Venice between mathematics and culture continues.

MICHELE EMMER

# Table of Contents

IX

# Homage to Mario Merz

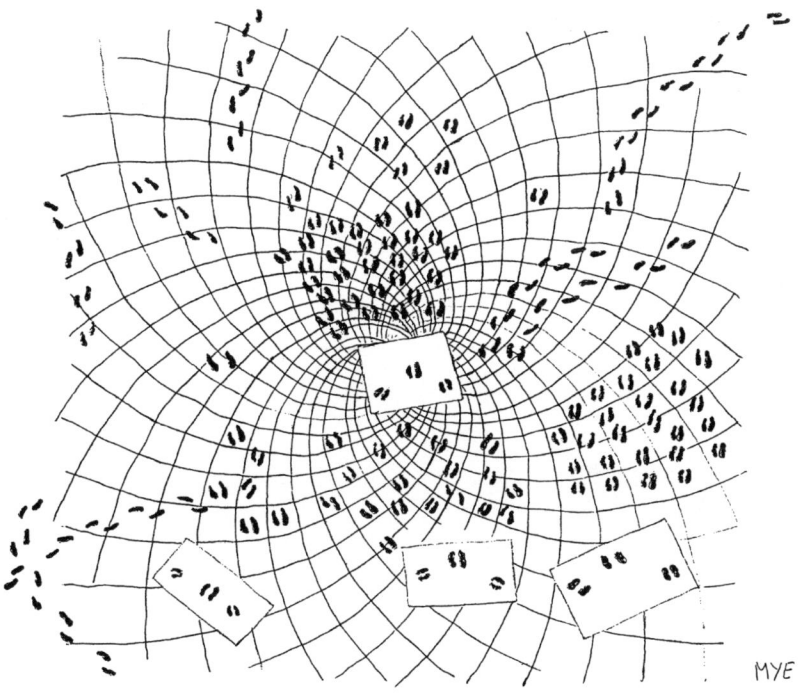

# The Eclipse*

Manuela Gandini

He plants his elbows on the desk. Marisa's grey hair falls like silver lianas. We all talk and argue together over a glass of red wine, only to find ourselves in the same territory, life, with a final laugh at the Banco in Milan, in the eighties.

Some time later we come out of the snail shell: Frank Lloyd Wright's spiral, where Mario is one of the very few living Italian artists to exhibit, in 1989, at the Guggenheim in New York. It was the physical manifestation of his interminable, energetic, spiral journey. The motorcycle ran along the walls, along with lizards, as in a climbing dance, and the neon lights glowed over the grey stacks of newspapers where the brief journey of contemporary man was concentrated.

But life does not stop at the highest peaks of the art system. The rarified air oxygenates thought and it goes on still further, quite a bit. The energy of blue neon is renovated in the metaphysics of inert objects; it was the light that spoke.

Doctor Merz, who in 1945 was in prison because of his political activities, begins to draw on letters, on bread and cheese, and becomes, with the friends of poor art and his tie with Christian Stein, one of the most important and revolutionary contemporary artists. He uses the poetry of matter and the mathematics of Fibonacci to represent the processes of growth in the organic world.

He died on the night of the eclipse, while the shadow of the moon covered half of the planet. The shadow carried him away, drawn by a "strange wind that bothers me", as he said in the afternoon. On the day after his death, Monday, 10 November 2003, the newspaper *La Stampa* put on the front page a photograph of the splendid igloo that he built for the city of Turin. At the same time, *La Repubblica* carried a photograph of the heap of debris from the terrorist attack at Riad, which was frighteningly similar to a devastated igloo. The two images, very similar, were each the opposite of the other, art and life, light and shadow, east and west, as if the two hemispheres were united in a single circular organism of creation and destruction.

3

---

* *Translated by Kim Williams*

# Merz and Fibonacci, Vital Proliferations in Mathematics and Contemporary Art*

Giovanni Maria Accame

The attention that Merz brought to the Fibonacci sequence by making it the centre-piece of a substantial and significant group of works beginning in 1970 and lasting through the last years of his work, is probably the best-known aspect of a relation-ship between mathematics and contemporary art which has been studied prevalently within (but also confined to) the context of historic abstraction and its successive de-velopments. In saying this I don't mean that those who have carefully read the art of the second half of the twentieth century, including the beginning of the twenty-first, have not noticed the presence of mathematics in fields that are diverse and also very distant from abstract painting, but rarely has this been taken into consideration by the critics. The chance to explicate the relationship between Merz and Fibonacci ne-cessarily leads me to delineate, at least summarily, the starting points for a reflection that could be, with justification, considerably expanded, precisely in the direction of an investigation into little-observed artistic intentions from this point of view, or even apparently adverse to a mathematical idea. This is the case of Merz himself and *arte povera* (poor art), of which Merz, who born in Milan in 1925 and died in 2003, was one of the leading figures. A clarification that is also necessary in order that the genesis of these works not be considered a bizarre anomaly, but rather of a more acute and pertinent intuition that could make a "poor" exponent the vehicle of a poetic that is tied to a dynamic and vital idea of nature.

Going back to the beginning of the 1960s and the artistic culture that was form-ing and rapidly asserting itself, what is immediately evident is the prevalence of a reaction to the long dominance of European informalism and of American ab-stract expressionism. Already by the end of the previous decade a clear and res-olute sign united various young artists on the two continents: monochromy. The monochromistic declension, intended as a detachment that is not without elements of continuity, above all on an existential level, as I prefer to see it, represents in any case a will to go beyond with regards to the gestuality and materiality that had char-acterised the anguish of informalism. Of the various lines of research that can be

---

* Translated by Kim Williams

delineated, especially in that with a minimalist and conceptual imprint we find an increasing number of artists who drew ideas for their work from mathematics. *Minimal Art* in the United States, as well as a widespread radical trend, often led to reflections on the essential structure of forms, materials and events. In more than one case the original cell, the internal order, is found in mathematics. An idea which, from a strictly disciplinary point of view, rarely ever in these cases offers an novel elaboration; the innovative aspect is obviously put forward by the interpretation given by the artist and in the work that springs from it.

I will limit myself to recalling a single artist, Sol LeWitt, as an example from among the most evident and relevant on the entire minimalist and conceptual panorama, but also, at the same time, vehicle of a model that is opposite to Merz. A model that appertains to the prevalent genre of connection between mathematics and art, in which, practically, the geometric figures and the subdivision of spaces reveals their origin immediately and formally. For quantity and quality of pertinent work LeWitt is certainly an obligatory point of reference for any itinerary whatsoever into the contemporary relationship between art and mathematics. Both in his plastic works and in those realised through wall drawings, the constructive motor is entrusted to progressions and developments that originate in mathematics and geometry. In LeWitt one can see a more articulated application and a greater complexity of solutions of the most significant constants of many of the relationships between artistic and mathematical work: repetition, modularity, iteration of variants. Beginning with the early years of the 1960s, the structures based on cubic models and executed in painted wood constitute the most evident contribution to the problem that interests us here. The artist himself is however concerned with making explicit that, on his part, there is not an explicit desire to face mathematical problems but rather to proceed along the way to a conceptualisation of art. Mathematics, in this sense, appears to make a considerable contribution to the reinforcement in art of the elaboration of ideas, as if to replace, as in effect it is in the intentions of conceptualism, attention and consideration of traditional manual techniques with an aesthetic evaluation of mutated structure from linguistics to logic, mathematics, etc. In his famous essay "Paragraphs on Conceptual Art", among other things LeWitt states:

> In conceptual art the idea or concept is the most important aspect of the work. When an artist uses a conceptual form of art, it means that all of the planning and decisions are made beforehand and the execution is a perfunctory affair. The idea becomes a machine that makes the art. [1]

This last statement, which cannot but recall Warhol's exclamation "I want to be a machine", takes to an extreme a particular notion of conceptualisation: the vision of its abstract mechanisation. The idea, which could be perceived as complexity and a plurality of solutions, even more so if it is developed in an artistic context, is rather understood in its function as a machine that automatically produces art. Less tied to geometry and to its objective realisation are the wall drawings by LeWitt, in which the mathematical and conceptual components produce an effect of absolute dematerialisation, conjoining and indicating with great effectiveness the measure and the poetics enclosed in an idea.

The opposite of this version of anti-expressionistic and micro-emotional conceptualisation is Mario Merz's work, who will develop a strong conceptual presence in his own linguistic maturation. For the young artists of *Arte povera*, and in particular for an older Merz, who by the mid-1970s had behind him significant painting experience, the new work that was developed in this period clambers over the individual sources that can be found in an analytic reading, to become that artistic phenomenon which had quickly achieved an extraordinary international success.

For Mario Merz the presence of nature and, even more, the sense of its generative force, are already found at the origin of the first artistic experiences. This is confirmed by paintings such as *Foglia* of 1952, *Seme del vento* and *Albero* of 1953, to cite only a few examples. When, in 1965, the passage to three-dimensional works occurs, and the use of neon begins, the idea of energy is not weakened, but reinforced. It is precisely neon, through a canvas, an umbrella, a bottle, a raincoat or other objects, which heightening Fontana's suggestion, giving body to a projection of light that cuts and, at the same time, regenerates.

Merz's encounter with the Fibonacci series is therefore logically consistent with his work. Certainly the artist deserves credit for having captured both the open and infinite dynamic aspects of application, as well as the extraordinary link between a mathematical idea and the actual progressions of nature and their potential evolutions.

As I have already mentioned, the distance that appears to exist between mathematical ideas and what *arte povera* has proposed to us through the works of its artists should not veil another essential aspect that belongs to more than one protagonist of this trend, constituting one of its distinctive characteristics. I am referring to the concept of primary energy, which distinguishes not only Merz, but which, in different ways, we find in evidence in Anselmo, Zorio and Penone, without forgetting the Romans Pascali and Kounellis. A relationship between the forces of nature and life and the generative creativity of numbers that was very clear to Merz, who in an interview in 1972, on the occasion of his exhibit at the Walker Art Center in Minneapolis, said:

> Man loves trees because he understands that they are part of an essential series of life. When a man has this kind of relationship with nature, he understands that he as well is part of a biological series. The Fibonacci series is natural. If I put a series of trees in an exhibit, you will have dead entities. But, in an exhibit, the Fibonacci numbers are alive, because men are like numbers in a series. People know that numbers are vital, because they can go forward into infinity, while objects are finite. Numbers are the vitality of the world. [2]

Merz, with a longer artistic history with respect to the other artists who would devote their lives to *arte povera*, found himself, in the mid-1970s, in a position that was particularly favourable for taking on a linguistic revolution, to the determination of which he himself contributed significantly. Indeed, he can free up all his creative impulses and further, thanks to his vast experience, have good control over them, and particularly, a notable awareness. It is not secondary that the two fundamental themes that were protracted in the installations of over thirty years, the igloo and the Fibonacci series, appear in 1968 and 1970, that is, in the very first years of the evolution that occurred in his work.

There is a Fibonacci Igloo, *Fibonacci Unit*, of 1970, among the less well known and reproduced, extraordinarily essential, that melds the constitutive structure of the igloo with the Fibonacci progression (Fig. 1).

If I recall correctly, this is the only work in which the formal elaboration of the hemisphere is completely determined by the numeric progression. Like a large spider, the igloo is delineated though a tubular structure in iron in which eight legs descend from the upper central part, their extension determined by the numeric progression, made visible by a series of joints that punctuate the individual units of measure that are the result of adding the two preceding ones. This work shows us clearly at least three important aspect of the artist's evaluation of the Fibonacci series. First of all, the constructive/generative nature inherent in this series, immediately grasped by Merz, one of whose fundamental poetic beliefs was the idea of expansion, of a new natural order in which biology and technology can find a synergetic motive in continuous growth. In the second place, the formal and spatial translatability of the numeric progression that, beyond the known and meaningful relationship with the spiral, can be extended infinitely into space. Finally, the symbolic value of the series in itself and the capacity of the numbers themselves to become signifying and symbolic elements of extraordinary incisiveness. This last characteristic is confirmed by the many works realised through mere dislocation in the context of neon numbers.

The *Progressione di Fibonacci* at New York's Guggenheim Museum of 1972 (Fig. 2), the installation realised for Minneapolis's Walker Art Center in 1972, that

**Fig. 1.** *Fibonacci Unit*, 1970. Copper, steel, marble, From the collection of the Kunstmuseum Wolfsburg

**Fig. 2.** *Progressione di Fibonacci*, 1971. Neon numbers

on the Torino's Mole Antonelliana in 1984 (Fig. 3), and the *Manica lunga da 1 a 987* at the Castle of Rivoli in 1990 are some of the major works, in terms of dimension as well, in which the Fibonacci series is made concrete solely through the presence of luminous numbers.

In these instances, the series does not activate its own formal elaboration, but superimposes itself on a space, or in the cases mentioned, on an architecture that is already defined, indicates a parallel but autonomous line of development. Merz has no intention of transforming the construction or of changing its meaning; if anything he wants to open a channel of communication, to add an itinerary, a perspective of thought that presents its own dynamic. To be sure, the combination results in other observations, other ideas, and this is part of the Fibonacci proliferation that enters life through nature and mathematics. In this sense nothing can be said to be definitive or immutable; the power of these numbers laden with energy acts on the imagination of those who are capable of imagining and know how to see different trajectories within a single reality. Regarding the concept of a plural reality, the artist himself says:

> The form of the numbers of the Fibonacci series is the form of the growth of many, many realities. In my work I love to use very simple mathematics. This series is the simplest series in the world. It is like counting, but it is a completely different way of comprehending the reality of mathematics. For me mathematics is an example of life, but not a reality in itself… The tenth Fibonacci number is more powerful than 10. A diagram of consecutive numbers is a straight line, but the diagram of Fibonacci is a curve that develops, like a spiral. The series

9

10

**Fig. 3.** *Il volo dei numeri* (Flight of numbers), 2000. Red neon numbers following the Fibonacci series. Photo: Paolo Pellion di Persano, Turin (see the section in colour)

has intense power. I believe that the reality of the world is like the Fibonacci series. [3]

This last statement confirms for us how much importance Merz attributed to this mathematical idea and that seems to coincide with the existence of the world itself. The energy of nature, its unstoppable becoming, and the exceptional dynamism that the series possesses and concretely expresses become the tool with which the artist indicates the intensity of what is happening. With *Tavole* of 1964 (Fig. 4) and other works on this same theme, the Fibonacci series is translated into objects that are very familiar to us, which the artist considers to be organically tied to man and his life environment: "the table is a piece of earth that rises up" [4], tables are for eating, working, playing, etc.

This is a relationship that Merz exalts, both in dimensional growth of the tables and in the number of the people sitting around them, all determined by the numeric progression. Another particular emphasis is found in the many works in which the Fibonacci series are combined with bundles of newspapers (Fig. 5).

**Fig. 4.** *Tavole con le zampe diventano tavoli* (Tables with Feet become Tables), 1974. Ink on canvas. From the collection of Kröller-Müller, Otterlo. Photo: Paolo Pellion di Persano, Turin

**Fig. 5.** *21 funzione di 8* (21 function of 8), 1971. Newspapers, glass, neon numbers

> I use newspapers because they are reproductions of words and thoughts (see [2]).

> … in New York you can see bundles of newspapers tied with a string. They are picked up and brought to the gallery … Newspapers interest me because there is an incredible unity in them. And they are all unread newspapers, something like a refusal of society [4].

It is difficult, however, not to associate the accumulation of bundles and bundles of newspapers with the idea of a flow of "everyday life" and thus with a dynamic of time tied to the series of the mathematician from Pisa. In the installation *610 funzione di 15* (615 function of 15) of 1971, *La natura è l'arte del numero* (Nature is the Art of Number) of 1976, and the more recent *Il fiume scorre* (The River Runs), the presence of the everyday, their adding up and stratification objectifies the flow of time that in reality always accompanies the works that we are dealing with here.

The profound connection between the Fibonacci series and nature, which so impassioned Merz, comprises the totality of our world and thus that space and that time that is so discernable in his works. Space and time conceived not only as containers, but as determining forces in how things happen, in the proceeding of that infinitely varied but substantially unitary flow that is life. Understood globally and dynamically to be in expansion, as happens in the spiral delineated by the Fibonacci numbers and as happens with the evocations that Merz's works continue to generate.

12

## Bibliography

[1]  S. LeWitt (1967) Paragraphs on Conceptual Art, *Artforum*, New York.

[2]  R. Koshalek (1972) Interview with Mario Merz, in: catalogue *Mario Merz*, Walker Art Center, Minneapolis.

[3]  B. Reise, L. Morris (1976) Eine Zahl ist ein Symbol für Wirklichkeit und Wachstum, *Kunstforum*, Mainz.

[4]  J.C. Ammann, S. Pagé (1981) Intervista a Mario Merz, in: catalogue *Mario Merz*, ARC, Paris – Kunsthalle, Basil.

# The Cinema According to Fibonacci*

Davide Ferrario

I was never very good in mathematics. It was a little bit my fault, and a little because in high school the teachers were always changing. Not to mention the particular moment in history (the seventies), which weren't conducive to unremitting
study … And yet the structural aspect of mathematics has always fascinated me, the
vague, almost miracle-working, perception that in numbers lay the key for explaining everything. So, in spite of my ignorance, I was always attracted to those theorems
that seemed to open the door to the understanding of how the world works. In fact,
I have always been convinced – even though I am an atheist – that nothing around
us (and maybe inside of us too) happens by chance. That's why I was easy prey to
the Fibonacci series.

I don't know when the first time was that I read about the unique properties of
the Fibonacci numbers. Not only the systematic elegance of the progression (each
number as the sum of the two before it), but also the fact that the further the series
goes, the closer it comes to the golden number, with all of the cabals associated with
it (speaking of which: did you know that the modern cinema's panoramic screen
was conceived on a ratio of 1:1.66, very close to the golden number?). More than
anything, I was fascinated by the suggestion (which still remains only that, even after
reading Mario Livio's beautiful book *The Golden Ratio*) that all – in numbers as in
nature, in architecture as in aesthetics – all, I repeat, seams to correspond to a general
rule of order and of development, connected in some way to the Fibonacci series. Of
course, I am not the only one who has fallen for this. Many artists seem to respond
with enthusiasm to this kind of oracle, making it the touchstone of their inspiration.
One of these was Mario Merz, who has threaded the Fibonacci numbers and spirals
in almost all of his works. Including, inevitably, Turin's Mole Antonelliana.

I think I looked at the installation (a vertical file of red neon numbers that follows the silhouette of the dome) for a couple of years before I ever thought seriously
about it. That happened, of course, when I decided to film *Dopo mezzanotte* (After
Midnight), a movie that takes place entirely inside the Mole. The movie's plot, for
those who haven't seen it, can be summed up briefly: it is the improbable love story

* *Translated by Kim Williams*

between the night watchman of the Mole, Martino, and Amanda, a 'bad girl' from the outskirts of Turin, who, fleeing from the police, hides out in the Mole. The movie, filmed without a proper script per se, was constructed day by day. At a certain point, shooting the Mole by night, I asked myself: and Merz's numbers? There they were, but we could have simply turned them off and ignored them. Or we could have considered them like an inevitable message from destiny and incorporate them into the Karma of the film. Obviously, I chose the second option.

And thus was born Martino's fixation for Fibonacci, a fixation that provided the occasion for one of the most surreal and poetic scenes of the film – not least because of the reference to the poor mathematician from Pisa in some contexts that were unforeseeable, such as the fight between Amanda and her stupid friend in the middle of the square in front of the city hall ... But under the surface I am convinced that the presence of the Fibonacci numbers in *Dopo mezzanotte* corresponds to something much deeper and more meaningful. The hope expressed by Martino that "in spite of everything there is a sense to the world" and that the profound significance of the 'golden' aspect of mathematics lies precisely in this, is in reality my wish as a filmmaker that the film itself, in spite of its eccentricities and originality, responds to a harmony that is deep and not random.

*Dopo mezzanotte* was a great, unexpected, success, in Italy as well as abroad (it sold in more than a hundred territories, from Burma to Rumania to the United States). It was as if the world over the public – even without a star in the cast or direction – found something in the film that speaks to a sense of beauty that cannot be categorised in terms of civilisation, but is genuinely universal. Like mathematics ...

Going around the world with the film I noticed that the audience's reaction always followed the same scheme: first a kind of disorientation mixed with curiosity (like this: what kind of story is this?), then – as the film gradually unfolds and makes sense – there comes a sense of awed wonder that is a little childlike, just like when someone explains the Fibonacci series to you. What was scattered and simply 'numerical' takes on the connotations of the revelation of an ordered vision whose very harmony engenders – let me say it – a kind of existential serenity. All things considered, the same thing that I felt when, after months of editing and many, many off-track digressions, I found the key to putting the material in order and making the film that you see today.

Speaking of film editing, here too I want to bring into play, even if this is not actually mathematics, quantum physics. The piece I want to talk about is long one, is a fantastic story. It comes from a most interesting book, *Conversations: Walter Murch and the Art of Editing Film* by Michael Ondaatje, the Canadian author of Dutch origin who wrote *The English Patient*, but the real star of the book is Walter Murch, a famous American editor, collaborator of Coppola and Lucas. The book is structured as a series of conversations on editing, of questions and answers. At one point Murch tells Ondaatje the story of "Quantum Twenty Questions", a stupendous game invented by John Wheeler, a scholar of quantum physics who was a student of Niels Bohr in the 1930s. Wheeler is the one who invented the term "black hole". He knows everything there is to know about the best physics of the twentieth century.

He's still living, and continues to teach and write. His game reflects the way in which the world is structured at the quantum level. Let's say that there are four players: Michael, Anthony, Walter and Aggie. From the point of view of one of the players – Michael, let's say – it appears to be the classic game of "Twenty questions", which you might call "Normal Twenty Questions". So Michael leaves the room, convinced that the other three players are looking around, choosing and agreeing on one object that he has to guess in less than twenty questions.

Normally, the game is based on a combination of acumen and luck: "No, it's not bigger than a breadbox", "No, it's not edible", and so on. But in the version invented by Wheeler, when Michael leaves the room, the other three players don't talk to each other at all. Instead, each of them chooses an object without telling the others what it is. Then Michael is called back into the room.

There is an incongruence between what Michael thinks the situation is and what it really is, that is, no one knows what object anyone else is thinking of. The game, though, proceeds the same way, and that's where the fun is.

Michael asks Walter: "Is the object bigger than a breadbox?" Walter, who has chosen, let's say, the alarm clock, answers, "No". Anthony, however, had chosen the couch, which is bigger than a breadbox. And since Michael is getting ready to ask Anthony the second question, Anthony has to hurry up and find another object in the room – a coffee cup! – that is smaller than a breadbox. So when Michael asks Anthony, "If I emptied my pockets, would the contents fit inside this object?", the answer is "Yes". The object that Aggie has chosen may be a little pumpkin carved for Halloween – smaller than a breadbox and big enough to hold Michael's keys and change, so when Michael asks her, for example, "Is it edible?", Aggie answers, "Yes". This is a big problem for Walter and Anthony, who have chosen objects that are not edible: now they have to choose something edible, big enough to hold the things from Michael's pockets, and smaller than a breadbox.

So what happens is that a complex vortex of decisions is created, a logical but unpredictable chain of conditions and solutions of "if …" and "then …". To finish successfully the game has to produce, in fewer than twenty questions, an object that satisfies all of the logical requisites: smaller than a breadbox, edible, large enough to hold keys and change, and so on. There are two possible outcomes: The game is successful, and it winds up that Michael, who is still convinced that he has played "Normal Twenty Questions". In reality, no has chosen object X, and Anthony, Walter and Aggie have had to sweat to go through invisible mental gymnastics, always just a step away from failure. Which is the other possible outcome. In fact, the game can fail miserably. After the fifteenth question, let's say, the sequence of questions can have generated a series of requisites of such complexity that there is no object in the room that can satisfy it. When Michael asks the sixteenth question, Anthony folds and confesses that he doesn't know how to answer, and Michael discovers that truth: they have been playing "Quantum Twenty Questions". According to Wheeler, the nature of perception and reality – at the quantum level, and maybe not only – is in some way similar to the dynamics of this game.

When Murch read about this game, he immediately associated it with what happens in film making. As he explains, there is a game that everyone agrees on, which

15

is the script, but during the making of the film so many variables come into play that everyone interprets the script in a way that is slightly different than the others. The director of photography arrives at his own conclusion, at which point, maybe, they tell him that a certain part should be played by Clark Gable, at which he thinks, "Gable? I didn't think he was right for the part. Now I have to rethink everything". Then the set designer changes the set a little bit, and the actor says to himself, "This is my apartment? Well, then I must be a different person than I thought I was; that means I have to change my interpretation". The camera operator, following him, has to choose a frame a little bit larger than he thought at first. At that point, with those images, even the editor is forced to change something, maybe giving the director an idea that results in the change of a line. When the costume designer finds out, he decides that the actor, in that scene, has to wear coarse socks. And so it goes. A film can be successful, spiralling around in itself until it reaches a final result that gives the impression of having been laid out from the beginning, even when in reality it is derived from a total shuffle.

On the other hand, the film can also fall to pieces. Emotional or logical inconsistencies can give rise to a question that nothing in the room – that is, the film – can answer. The most glaring example of this is the choice of the wrong actor, which poses a problem of consistency with all the rest. The film, however, can also fail for much more subtle reasons: killed by thousands of cuts, by interference from the studios, by bad weather, by the fact that the producer is getting a divorce, etc. All of these things turn out to be inscribed in the body of the film in the most complex ways. Sometimes the effects are good, and the film is enriched. Sometimes the effects are bad, and the film aborts, it is finished but never released; or maybe it is sent into cinemas with fatal flaws that only lead to bad reviews. The comparison between making a film and Wheeler's game helps us to answer the eternal question: what were they thinking of when they made that film? How could they have believed that it could have worked?

No one commits to producing a film that can't be released, but the game of the film can raise questions for its creators that, in the end, they don't know how to answer, and the film ends badly.

I have to say, as a cinematographer, that I have never read truer words about the essence of film making. Further on, Murch goes even further and provokingly maintains that it a film should be edited using *I Ching*. Each toss of the coin (which is not a kind of random horoscope, but responds to a more 'oriental' kind of approach to the calculation of probability) could provide the key to the way in which a scene is attached to another. It seems crazy and just a pretext, but how can you not see in this arrangement an attitude inspired by the positive principle expressed by Martino: "All told, the world (and cinema, I have to add!) has a meaning …"

From a philosophical point of view, the basic principle is that nothing is ever invented, at most it is discovered. This is exactly what I think when I make a film. I never believe that I am inventing something, I am profoundly convinced that it was already all there, it only had to be found. My job, my talent, lies in finding the way of deciphering the road to get there (and often that way is not at all logical, but follows the models of quantum physics described by Wheeler's game). All this has an

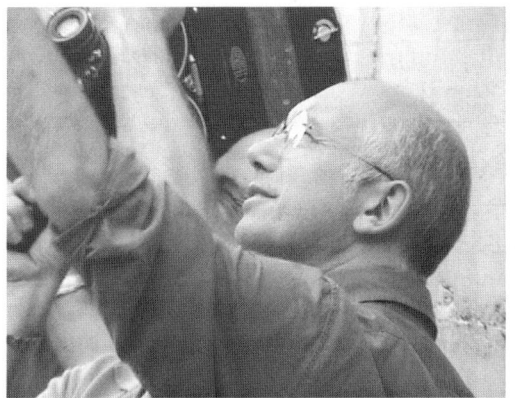

**Fig. 1.** Davide Ferrario on the set

**Fig. 2.** The film's leading actors, Giorgio Pasotti and Francesca Inaudi, in the Mole Antonelliana

**Fig. 3.** The film's third protagonist, Fabio Troiano

**Fig. 4.** Francesca Inaudi, and in the background, the outskirts of Turin

obvious correspondence with mathematics, where numbers are set out like a "text" that is already written and inalterable, behind which are hidden revelations that are just waiting to be discovered … So, I think that a film, at the moment when it is first conceived, is "already made". What needs to be found is the demonstration – in a mathematical sense – that, by means of all the adventures and vicissitudes of a film, proves the theorem.

*Dopo mezzanotte* presents itself then as a particularly interesting case of this theory because the theory is incorporated inside the plot itself, turning it into a genuine narrative mechanism. On the other hand, all the parts that relate to the sentimental plot of the film correspond to the Fibonacci numbers: 1, 2, 3. One, each single element involved. Two, the couples that are formed in the amorous 'patrol'. Three, the triangle, à la Jules et Jim, that is created at a certain point. (Let's not forget that Fibonacci's proof effectively started with a problem about rabbits mating …).

The definitive confirmation that all of what I have said is not pure raving came to me when I had opened the season for the cinema awards. In the midst of the general recognition of the film's value for various reasons, from direction to the actors to the set design, what received the most recognition (Nastro d'Argento, David di Donatello, Diamanti del Cinema Italiano, etc.) was the screenplay, which they even wanted to publish. What a shame that – as I said – a screenplay never existed! I read this fact as the counterproof that if a key is found to 'read the numbers' of the film, even film professions confuse the result with the production of a pre-existing and pre-ordered project.

To conclude, I think that cinema is both a work of 'decimal' precision and absolute, chaotic creative freedom. Just as, I like to think – and hope –, mathematics is.

# Mathematics and Images

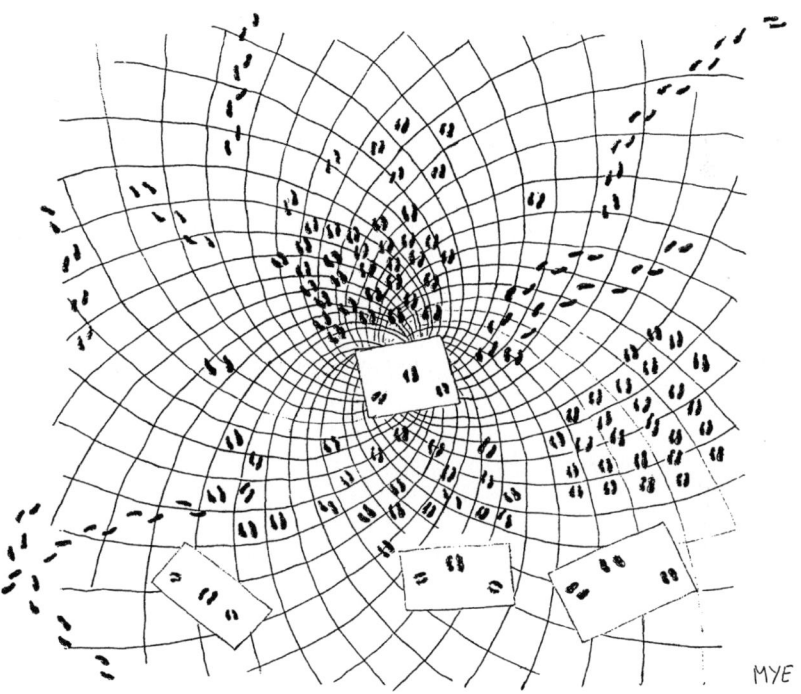

# PDEs, Images and Videotapes*

Maurizio Falcone

## Introduction

We all live in a world made up of images, films, cameras, and television news, but few know that mathematics has become a very useful instrument in the treatment of images and that many of the images that we see have been cleaned, filtered, and corrected by means of some software that uses advanced mathematical tools. In the field of image treatment, mathematical models burst onto the scene only about fifteen years ago, and since then, the development of increasingly sophisticated mathematical techniques for image manipulation has become widespread, thanks to the evolution of computers as well, which by now allow whoever owns a computer to manipulate the images from his camera (digital, of course) or to assemble a film (digital, again) filmed with his video camera.

We will try to understand better what these techniques are based on and which mathematical models are used in this field. The first thing to do is to give a mathematical definition of what an "image" is. A first, simple definition could be the following:

**Definition 1.** *An image is a rectangle of coloured points.*

In truth, this definition captures only some of the characteristics of an image: the fact that it is rectangular and that each point of the image is coloured. We do well to add another, less evident characteristic: an image is constituted of a finite number of points, called pixels.

This aspect is not clear when we look at a photograph, simply because the number of points on the photographic paper on which it is printed is so high that it appears to be infinite (the image is continuous, and does not appear grainy). But try to stop the image in a VHS cassette on your television and you will find that it is really made of many points arranged in rows and columns. The quality of television screens is actually much inferior to the grain of photographic paper (a few thousand points for a television screen and a few million points for photographic paper). The

---

* *Translated by Kim Williams*

reason why we are not very aware of the low quality of the images on our televisions is that our retina sees a sequence of images (the frequency of a film is twenty-four photograms per second) and the low-quality images are superimposed on our retina, giving the impression of continuity of shapes and movement.

We are then able to give a more precise definition of an image, and for the sake of simplicity, we will give the definition of an image that is commonly called a "black-and-white" image, even if, to be exact, we are dealing with an image in grey levels. There are usually 256 grey levels and by convention they go from 0 (black) to 255 (white).

**Definition 2.** *An image is a rectangular table of integer numbers between 0 and 255.*

In mathematics a rectangular table of numbers with $M$ rows and $N$ columns is called matrix $M \times N$. Each element of the matrix/table is easily identifiable beginning with two integer indices that identify the row and column to which it belongs. For example, the element $p_{ij}$ is the element on row $i$ and on column $j$, and the value $p_{ij}$ will be between 0 and 255 (Fig. 2).

**Definition 3.** *We will call an image the matrix I, $M \times N$, whose elements $I_{ij}$ are all integers comprised between 0 and 255, that is, $0 \le I_{ij} \le 255$ for $i = 1, \ldots, M, j = 1, \ldots, N$.*

What about a colour image? The simplest way to define it is to take advantage of the fact that each colour can be obtained as a combination of three fundamental

**Fig. 1.** A colour image (see the section in colour) and one in grey levels

$$\begin{bmatrix} 0 & 0 & 8 & 15 & \ldots & 21 & 33 & 32 & 32 & 31 \\ 0 & 0 & 0 & 42 & \ldots & 23 & 35 & 32 & 32 & 31 \\ \ldots & & \ldots & & \ldots & & & & \ldots \\ 0 & 0 & 52 & 10 & \ldots & 230 & 35 & 38 & 31 & 31 \\ 0 & 0 & 0 & 80 & \ldots & 130 & 35 & 38 & 31 & 31 \end{bmatrix}$$

**Fig. 2.** An image in grey levels is a $M \times N$ matrix

**Fig. 3.** The three RGB channels of an image of the Grand Canal (see the colour section)

colours. In the RGB system, commonly used in televisions and digital cameras, the three colours are red (R), Green (G), and blue (B). By choosing a combination of these three colours we can obtain all colours. For example, 50% of red mixed with 30% of green and 60% of blue produces a violet. We could therefore obtain a colour image by simply superimposing the three RGB channels. Each channel corresponds, as in the case of the grey-scale image, to a matrix $M \times N$ in which the element $ij$ is the value of the tone of the corresponding channel. Thus, if we indicate with R, G, and B the three matrices that correspond to the three channels, values $r_{ij}$, $g_{ij}$, and $b_{ij}$ correspond to the tone of red, green and blue of the point in the location $ij$ of the image, and each of these values is between 0 and 255.

Figure 3 shows the three RGB channels of the colour photograph of the Grand Canal shown in Fig. 1.

We will see how mathematical models for a single image can be used to deal with and resolve some problems in the field of image treatment, but first we will introduce some of these problems.

23

## Filtering

This is the problem of eliminating disturbances from an image (Fig. 4). These disturbances (so-called "noise") can be caused by several factors: disturbances of transmission/reception (as happens, for example, with images transmitted via satellite or in telecommunications), disturbances in the construction of an image (for example, due to a dirty lens or an imperfection in the system that reads the image). Each of these disturbances has its own characteristics, which can be described in mathematical terms; the goal is to eliminate all disturbances to obtain a sharp image.

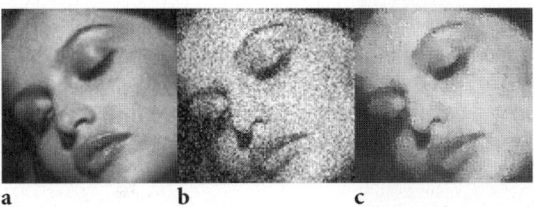

a    b    c

**Fig. 4.** A typical filtering test: **a** original picture, **b** original with Gaussian noise 10%, **c** filtered picture (see the section in colour)

## Segmentation

The problem of segmentation occurs when the borders of the objects represented in the image need to be determined accurately (Fig. 5a). In this problem the borders are identified as the zones where there is a strong variation in the grey level or colour (in the case in which each object is associated with a single colour). Segmentation has many applications and is often an obligatory step towards resolving other problems *only* on some parts of the image. For example, in remote sensing once the objects have been identified it is possible to count them to determine the number of buildings, trees or cars present in a certain area.

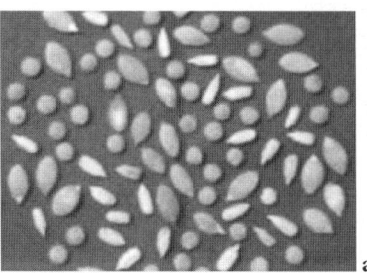 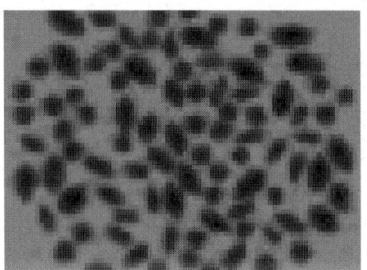

**Fig. 5a,b.** Segmentation of grains: **a** original; **b** segmented photo

## 3D Reconstruction

In this problem we want to reconstruct the surface of the object represented in the image (Fig. 6a). In order to do this, it is first necessary to isolate the object that interests us, possibly by means of a segmentation, and to use the information about the grey levels contained in the image to reconstruct the surface. For this reason, the problem is known in the literature by the name "Shape-from-Shading" (SFS). It is a difficult problem, whose applications are potentially extremely vast and go from security (the automatic recognition of a person) to remote sensing (for the construction of maps of the surface of the earth or other planets).

 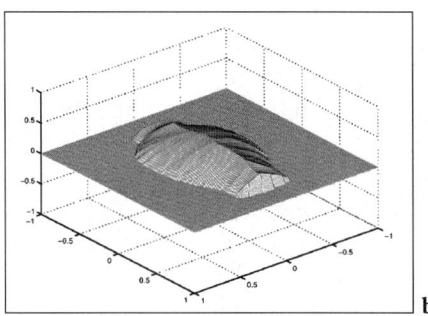

**Fig. 6a,b.** 3D reconstruction of a vase: **a** original; **b** surface

## Zoom

We want to be able to enlarge an image without losing definition of the objects and without a loss of focus (Fig. 7a,b). The main difficulty is related to the fact of enlarging an image that has few pixels so that it becomes a image with many pixels without having any other information about the objects represented in the image. In this problem as well it is necessary to keep the borders of the objects sharp during the process of zooming.

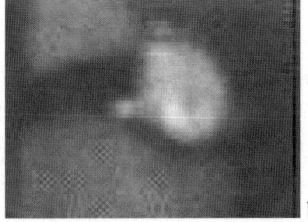

a    b

**Fig. 7a,b.** The enlargement of a detail (the watch) leads to a loss of definition

The solution to problems such as these can be applied to many fields, ranging from biomedical research (sonograms, CAT scans, electron microscopy) to space research (telescopic images, radar clinometry) (Fig. 8).

How are these problems resolved in the area of image treatment? For many years the discrete model of an image was the most used, but for some years now we have begun to use more refined mathematical models that are based on partial derivative equations. In some of these models the step from an initial image to a treated image (cleaned, filtered, segmented) is described through the study of the process of evolution of the function $I(t, x, y)$. The mathematical description of this process ties the time variation of the image (that is, the temporal derivative $I_t(t, x, y)$ of the intensity of the grey level at every point and every instance of time) to its local characteristics, described by its spatial derivatives (first and/or second order). Even though the speed of this evolution depends on the model and on the result that one wants to obtain, the presence of the derivative of the function that is being studied requires a certain regularity in the solution, which must at least be continuous. This require-

25

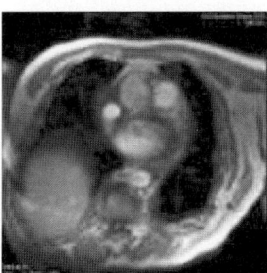

**Fig. 8.** Biological and astronomical applications

ment leads to abandon, in this phase, the representation of an image by means of a matrix of values (a discrete model) in order to shift to a continuous model in which the grey levels are infinite and vary between 0 (black) and 1 (white). In the continuous model the intensity of the levels of grey at time $t$ of the evolution is defined at all points $x$ of the rectangle corresponding to the image, thus $I: Q \times [0, T] \to [0, 1]$. To resolve the problem of evolution, generally rather complex, it is necessary to use methods of approximation and to introduce steps of discretization ($\Delta t$ for time, $\Delta x$, $\Delta y$ for space). Thus the effective numerical solution of the model becomes discrete again and the representation of the image is once again a matrix with the same number of points of the starting image. A schematic representation of the above process is the following.

$$
\begin{array}{ccccc}
\text{DISCRETE} & & \text{CONTINUOUS} & & \text{DISCRETIZATION} \\
\text{MODEL} & \Rightarrow & \text{MODEL} & \Rightarrow & \Delta x, \Delta y \\
\text{Matrix } I(i, j) & & \text{Function } I(x, y) & & \text{Matrix } I(i, j) \\
0 \leq I \leq 255 & & I : Q \to [0, 1] & & 0 \leq I \leq 255
\end{array}
$$

But why is it useful to consider a continuous model if the image is described by a discrete structure that corresponds, as we have just seen, to a matrix? The main reason is that often it is simpler to describe a phenomenon by means of a continuous model in which we can use the instruments of mathematical analysis to arrive at a representation that is more concise and precise, rather than describing the evolution of each point of the discrete image. The continuous model often uses the theoretical developments related to non-linear differential equations and their numerical treatment. For example, it is possible to determine the conditions under which there is a unique solution to the problem, what their characteristics are during the evolution over time, what its maximum and minimum values are for the levels of grey $I$, and, possibly, for their derivatives.

In constructing the model, however, we encounter some difficulties. The first is that the images contain a great amount of information and the model has to be able to select the specific information that is relevant for the solution to the problem at hand. The second regards the fact that the borders of the objects have vertexes and edges and this poses the problem of the calculation of solutions that are not regular (that is, that don't have derivatives). Finally, the images contain a great mass of data: for example, a small image ($512 \times 512$ pixels) in grey levels occupies 262 Kb. Obtaining results in reasonably short time requires optimised and fast algorithms.

We will concentrate on two classical problems of image treatment in order to better explain what kind of mathematical models are used.

## 3D Reconstruction (SFS)

In this model there is no evolution over time. Given an image $I(x, y)$ in grey levels, we wish to reconstruct the surface $z = u(x, y)$ that corresponds to it (Fig. 9).

The inverse problem is generally badly expressed, in the sense that a unique solution does not exist and small perturbations in the data can result in great variations

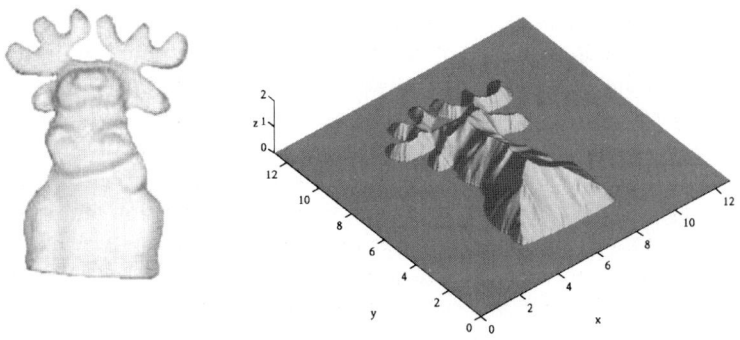

**Fig. 9.** Image and surface reconstructed in a SFS problem

in the solution. To simplify the problem and arrive at a differential model some hypotheses are introduced [7]:

H1. the light source is located at infinity in direction $\omega$; for example, $\omega = (0, 0, 1)$ in the case of a vertical light source;

H2. the surface of the object has properties of uniform light reflection (a Lambertian surface);

H3. the distance of the objective is large with respect to the object (so that perspective deformation is eliminated).

Under these hypotheses it is possible to describe in a simple way the relationship between the light measured at a point of the image, the normal $n(x, y)$ to the surface at point $P = (x, y, u(x, y))$ and the direction of light $\omega$:

$$I(x, y) = \gamma n(x, y) \cdot \omega . \tag{1}$$

Equation (1) is called the *radiance equation* and describes the dependence, in this model, of the light measured on the image in grey levels solely on the scalar product of n and $\omega$ and from the parameter $\gamma$, which describes the physical property of surface reflection. Parameter $\gamma$ is called the *albedo* and is presumed to be known (for simplicity in what follows we will set it equal to 1). If the surface is a graph $z = u(x, y)$, the normal to the surface at the point corresponding to $(x, y)$ is given by:

$$n(x, y) = \frac{1}{\sqrt{1 + |\nabla u|^2}}(-u_x, -u_y, 1) \tag{2}$$

where $u_x = \partial u / \partial x$, $u_y = \partial u / \partial y$ are the partial derivatives and $\nabla u \equiv (u_x, u_y)$ is the gradient of $u(x, y)$. In the case of a light source placed vertically $\omega = (0, 0, 1)$ as a particular case of (1) the following *eiconal* equation is obtained:

$$|\nabla u(x, y)| = \sqrt{\frac{1 - I^2(x, y)}{I^2(x, y)}} . \tag{3}$$

27

We can see that in the points of maximum brightness the known term becomes null, making the problem more difficult. In fact, in this case there is no unique solution, even when the value of $u$ on the edge of the image is imposed, as the following example shows. Considering, in $R^1$, the function $z = f_1(x) = -x^2 - 1$ in the interval $Q = [-1, 1]$, at the extremes of the interval the function equals 0. Any function $f_\alpha$ whatsoever that is obtained by reversing the function $f_1$ with respect to a horizontal axis $y = \alpha$, with $\alpha \in (0, 1)$ will again verify the equation in almost all points of the interval, as well as verify the boundary condition $f_\alpha(-1) = f_\alpha(1) = 0$. There are therefore an infinite number of solutions, depending on the value of parameter $\alpha$. This ambiguity can be explained by the difficulty in distinguishing, in the hypothesis of the model, a concave form (a hill) from a convex form (a volcano), which is known in the literature as the name "concave/convex ambiguity" in the SFS problem.

This example suggests two remarks. The first is that it is necessary to introduce additional information on the surface in order to be able to determine it uniquely. Various proposals have been put forth: the concavity/convexity of the solution can be fixed, or the heights at the points of maximum brightness can be fixed, or again a maximal solution can be defined (one larger than all the others) and selected from within the family of solutions $f_\alpha$. The second remark regards the fact that the functions $f_\alpha(x)$ do not verify the equation at all points of the interval but only where the solution can be derived, and thus where the derivative $u_x$ is well defined. Since there is an obvious interest in accepting these solutions in a weak sense as well (objects have corners!), it is necessary to give meaning to the solution even when the surface is not regular, and to select a unique solution. In this context played a very important role the concept of the solution in the "viscosity sense" introduced by M. Crandall and P.L. Lions in 1984 [3] (for the development of the theory see [1]).

In the last few years many researchers have studied the SFS problem in the attempt to eliminate some of the hypotheses and open the way to more realistic applications. Some recent developments are the elimination of hypothesis H1 and hypothesis H3 for treating perspective deformation in the model (Fig. 10). This problem is called "perspective SFS" (see [4], [5], and [11]).

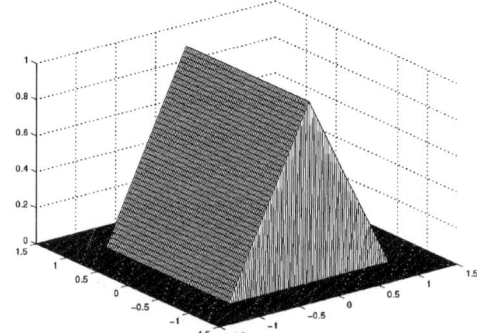

**Fig. 10.** Image and reconstructed surface in the perspective SFS problem

Other extensions concern the presence of zones of black shadow in the image (as happens in the case of an oblique light source [6]). In any case, the most complex problem remains open: how can hypothesis H2 be eliminated in the model?

## Segmentation

As we have seen, the goal is to determine the borders of the objects in an image. This result can be obtained with various techniques, for example, with the techniques of calculus of variations described in [9]. We will describe a recent technique that uses the evolution of an initial curve (a circle, for example) to determine the border of the objects. This technique, called *active contours*, is based on recent developments in the study of the evolution of fronts [2]. Let us see what this means.

To explain the evolution of a front we will take a problem of combustion, where it is relatively simple to define the mathematical objects that we wish to describe.

Let us consider a closed curve $\Gamma_0$ in the plane. This curve separates two zones: an *inner region* $\Omega^-$, which we will call the burned zone, and an unburned *outer region* $\Omega^+$. The curve $\Gamma_0$ represents the front at the initial time, which in the combustion model is the front of the flame (you can think of a dry field in which a fire is moving starting from the curve $\Gamma_0$). The speed of propagation of the front is directed along the exterior normal direction of the curve (thus the front tends to broaden itself by burning the outer zone as well) and depends on the physical characteristics of region $\Omega^+$. These characteristics can vary from point to point, and can give rise to different evolutions starting from the same initial curve.

The problem consists in determining the position of the front (that is, of the line that separates the burned zone from the unburned zone) $\Gamma_t$ at all successive times, that is, for all positive values of $t$. This problem presents various difficulties: the burned zone, for example, might be composed of diverse pieces (imagine burning the field in various points that are far apart) and with the evolution that we have described the fronts corresponding to the evolution of each of the initial burned zones might meet at a certain time to then continue as a single front (Fig. 11). This phenomenon is called *topological change*, since several fronts merge into a single one (Fig. 12).

The most used methods to describe these phenomena is the *level set method*, which takes its name from the fact that the initial curve $\Gamma_0$ in the plane is represented as the curve at 0-level of a function $z = u_0(x, y)$ chosen such that:

$$\begin{cases} u_0(x, y) < 0 & \text{in the inner region } \Omega^- \\ u_0(x, y) = 0 & \text{on } \Gamma \\ u_0(x, y) > 0 & \text{in the outer region } \Omega^+ \end{cases}$$

This method, introduced by Osher and Sethian [12], was very successful (see [13,14] for other developments and applications to images), and allows the determination of a solution to the problem of the evolution of a front by simply calculating the solution of the problem of evolution:

$$\begin{cases} u_t + v(x, y) |\nabla u(x, y, t)| = 0 & \text{in } Q \times (0, T) \\ u(x, y, t) = u_0(x, y) & \text{in } Q \end{cases}$$

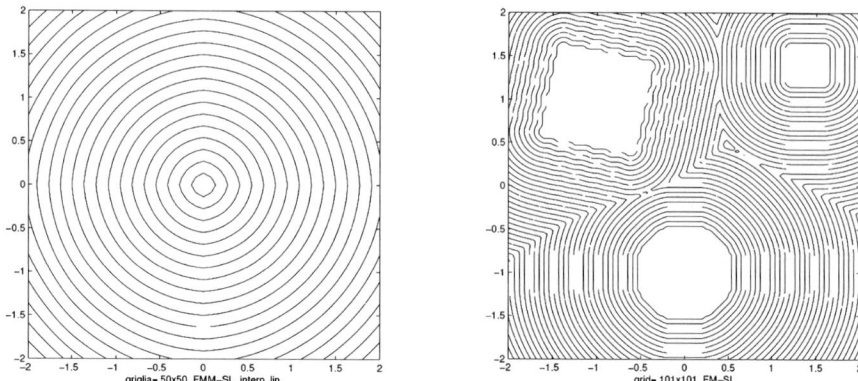

**Fig. 11.** Various types of evolution beginning from one or more initial fronts

where $v$ is a scalar function indicating the velocity in the normal external direction at point $(x, y)$. In fact, it has been proven that the 0–level curve of the solution $u(x, y, t)$ of the problem of evolution gives us the front $\Gamma_t$ at time $t$, even when topological changes are present and until the front is completely contained in $Q$.

In the applications to images, presuming the choice of the initial curve within the interior of the object (as in Fig. 12), the velocity $v$ of the curve still follows the exterior normal direction and is defined such that it is inversely proportional to the variations of the gray levels. The reason is rather obvious: the border of an object in an image is precisely the zone where an abrupt change in gray levels can be observed. Thus for segmentation it is natural to choose the velocity:

$$v(x, y) = \left(1 + |\nabla I_0(x, y)|^p\right)^{-1} \quad \text{with} \;\; p \geq 1$$

where $I_0$ is the function that describes the gray levels of an image and $\nabla I_0$ is the vector of its partial derivative, that is:

$$\nabla I_0(x, y) = \left(\frac{\partial I_0(x, y)}{\partial x}, \frac{\partial I_0(x, y)}{\partial y}\right).$$

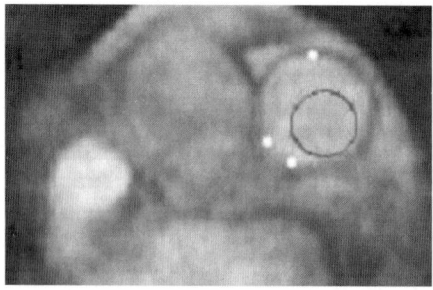 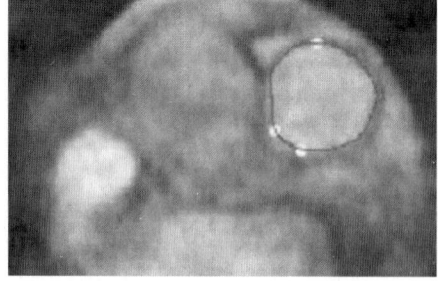

**Fig. 12.** Propagation of a front for determining the border in a sonogram

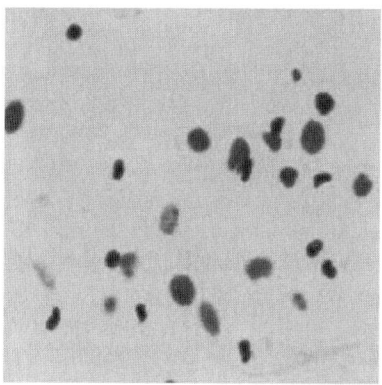 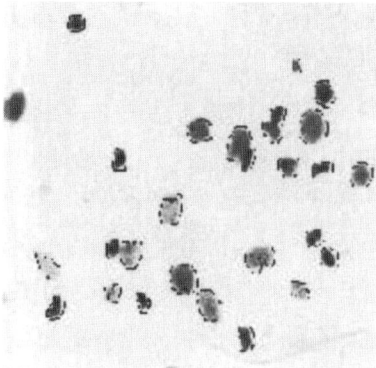

**Fig. 13.** Segmentation of an image of bacteria (electron microscopy)

Since in the image there can be several objects, the curve has to be able of being divided into various parts (topological change) and double points and singularities can be created.

This approach has allowed the problem of segmentation to be solved automatically for biomedical images (from electron microscopes (Fig. 13), sonograms, X-rays) as well as for satellite images. In these contexts the problem is more complex because of the presence of strong disturbances in the image (the "noise" we mentioned earlier).

The success obtained in the segmentation of individual images has suggested the extension of this technique to the segmentation of movies. That is, the goal is to determine the border of an object in the first frame and then to follow it in successive frames. Obviously, this problem is much more complex, because it is necessary as well to estimate the direction in which the object will move from one frame to the next (*problem of optical flow*). Nonetheless, the results of experiments are rather impressive and can be found on the websites of some research centres (see, for example, [10, 14]).

The treatment of images has become one of the more active area of applied research for mathematicians, since in this field non-linear differential and variational models developed in the last few years have turned out to be very useful. It is interesting to note that new problems have emerged from the interaction with industry. They demand new theories and new models that will be probably used to design new electronic gadgets.

## Bibliography

[1]  G. Barles (1994) *Viscosity solutions of Hamilton–Jacobi equations*, Springer-Verlag.
[2]  T. Chan, L. Vese (2002) *Active contour and segmentation models using geometric PDE's for medical imaging*, Geometric methods in bio-medical image processing, pp. 63–75, Math. Vis., Springer, Berlin.

31

[3] M. Crandall, P.L. Lions (1983) Viscosity solutions of Hamilton-Jacobi equations, *Trans. Amer. Math. Soc.* 277, pp. 1–42.

[4] E. Cristiani, M. Falcone, A. Seghino (2005) Numerical Solution of the Perspective Shape-from-Shading Problem, *Proceedings of Control Systems: Theory, Numerics and Applications*, 30 March–1 April 2005, Roma. PoS (CSTNA2005) 008, http://pos.sissa.it/

[5] Website of J.D. Durou (IRIT),
http://www.irit.fr/~Jean-Denis.Durou/

[6] M. Falcone, M. Sagona, A. Seghini (2001) A global algorithm for the Shape-from-Shading problem with black shadows, in F. Brezzi, A. Buffa, S. Corsaro, A. Murli (eds) *Numerical Mathematics and Advanced Applications* – ENUMATH 2001, Springer-Verlag, pp. 503–512.

[7] B.K.P. Horn, M.J. Brooks (1989) *Shape from Shading*, The MIT Press.

[8] S. Osher, R.P. Fedkiw (2003) Level Set Methods and Dynamic Implicit Surfaces, *Applied Mathematics Sciences* 153, Springer.

[9] J.M. Morel, S. Solimini (1995) *Variational methods in image segmentation*, Birkhäuser.

[10] Website of N. Paragios (INRIA),
http://www-sop.inria.fr/robotvis/personnei/nparagio/demos/

[11] E. Prados, O. Faugeras (2003) *Perspective Shape-from-Shading and viscosity solutions*, IEEE, Proceedings of ICCV 2003, pp. 826–831.

[12] S. Osher, J. Sethian (1988) Fronts propagating with curvature-dependent speed: algorithms based on Hamilton-Jacobi formulations, *Journal of Computational Physics* 79, pp. 12–49.

[13] J. Sethian (1999) *Level set methods and fast marching methods*, Cambridge University Press.

[14] Website of "Image Processing" research group at UCLA (USA)
http://www.math.ucla.edu/~imagers/

# Mathematics in the Air with Solar Impulse

Alfio Quarteroni, Gilles Fourestey, Nicola Parolini,
Christophe Prud'homme, and Gianluigi Rozza

## The Flight Dream

Ever since the drawing of flying machines by Leonardo Da Vinci (1452–1519) flight has been a long-lasting dream of Mankind. After one century of aviation history (since the *première* of the Wright brothers on December 17th, 1903) and the development of commercial flights all over the world, a new era in aviation is opening with impressive new challenges and missions, and also innovative concepts for flying machines. A recent mission (March 4th, 2005) has seen the *Virgin Atlantic Global Flyer*, piloted by the aviator and record breaker Steve Fossett, successfully landing at Salina, Kansas and completing the first solo, non-stop round-the-world airplane flight (67 h 2 m 38 s) with a single engine turbofan aircraft [1].

Another challenging program has been carried out by NASA since 1999 with *Helios Prototype*, a unique solar-powered experimental lightweight flying wing. An unofficial record altitude for non-rocket-powered aircraft of 96,863 feet was achieved in 2001.

In this spirit, Bertrand Piccard has proposed in 2003 the first round-the-world solar airplane flight which aims to promote a new ethos of sustainable development through an exciting new adventure. *Solar Impulse*, the name given to the project, will re-create all the great firsts in aviation history, culminating with a flight around the world, using only renewable forms of energy and without generating any polluting emissions (see Fig. 1).

The project combines technological innovation, human adventure and respect for the environment, drawing public attention to the essential changes that are necessary to ensure future energy resources and the ecological balance of our planet, and reinforcing the idea that technology can work hand-in-hand with the objectives of endurableness.

So far, little has been achieved by solar-powered airplanes. They cannot remain airborne for more than a dozen hours due to limited energy resources. To help the Solar Impulse team, the École Polytechnique Fédérale de Lausanne (EPFL), the official scientific advisor of the project, will draw upon intellectual and scientific re-

**Fig. 1.** Solar Impulse: a project to achieve the first round-the-world solar airplane flight (see the section in colour)

sources from more than ten diverse research domains to focus on several technological challenges: ultralight materials, novel energy storage and retrieval systems, and new types of human-machine interfaces [2].

Mathematical modeling, scientific computing, numerical simulation (CFD, mesh generation, geometrical reconstruction) and multi-objective optimization techniques play an important role in this multidisciplinary project concerning flight strategies evaluation as well as aero and structural dynamics analysis.

## Exploration, a Tradition of the Piccard Family

Pioneering exploration is a tradition in the Piccard family from the generation of Auguste to Bertrand (his grandson) without forgetting Jacques (Auguste's son).

Auguste Piccard, born in 1884 in Basel, Switzerland, was professor of physics at the Swiss Institute of Technology in Zurich and then at the University of Brussels. A friend of Albert Einstein and Marie Curie, he contributed to modern aviation and space exploration by inventing the pressurized cabin and the stratospheric balloon. He made the first ascents into the stratosphere in 1931 and 1932 to study cosmic rays, reaching heights of 15,781 and 16,201 meters respectively. He became the first man to observe the curvature of the Earth with his own eyes. Applying the principle of his stratospheric balloon to the exploration of the deepest oceans, he built a revolutionary submarine, which he named *Bathyscaphe*. Diving with his son Jacques to 3,150 meters in 1953, he became the man of both extremes: having flown the highest and dived the deepest.

Jacques Piccard was born in Brussels in 1922. After his initial studies in economics, his connection with the business world enabled him to raise funds for his father's second bathyscaphe. Jacques then changed his career and worked with his father to build what was to become the bathyscaphe *Trieste*. Diving with Auguste, he broke numerous records before himself capturing the World Record for the deepest ever dive, 7 miles down to the bottom of the Marianas Trench. After his

**Fig. 2.** The Piccard family: *from left*, Auguste, Jacques and Bertrand

father's death, he continued the family mission constructing mesoscaphes – submersibles designed for medium depths. The first was the famous *Auguste Piccard*, the world's first passenger submarine. At the 1964 Swiss National Exhibition in Lausanne, it took 33,000 tourists to the depths of Lake Geneva. Next came the *Ben Franklin*, with which Jacques explored the Gulf Stream in 1969, drifting 3,000 km in a dive that lasted a month; and finally the *F.-A. Forel*, an easily transportable pocket-submersible, in which Jacques made more than 2,000 scientific and educational dives in European lakes and in the Mediterranean Sea.

Bertrand Piccard, born in 1958 in Lausanne and long interested in studying human behavior under extreme conditions, was one of the pioneers of hang-glider and micro-light flying in the 1970s, becoming European champion in hang-glider aerobatics in 1985. Introduced to ballooning, he won the first transatlantic balloon race with Wim Verstraeten (the *1992 Chrysler Challenge*). He then launched the *Breitling Orbiter* project to fly around the world, and was captain of all three attempts. With the British aeronaut, Brian Jones, he accomplished the first ever non-stop flight around the world in a balloon, which became also the longest flight in both distance and duration in the history of aviation (1–20 March 1999: from Switzerland (Châteaux d'Œx) to Egypt) [3].

35

## Easing Our Way Out of the Fossil Age

Today, the main source of energy is oil. Its by-products are found virtually everywhere: fuel for cars, planes and heating devices, roads, plastics component ... it is the foundation of the global economy. However, oil resources are not limitless. Specialists consider that the total oil resource of the Earth is about 3,000 billions of barrels: 1,000 billions already depleted, 1,000 billions located, 1,000 billions left to be found. The Hubbert curve is often used in order to describe the world trend of oil production. This gaussian curve will reach its maximum when half of the oil resource will have been exploited. There are several estimations of when this will happen, ranging

**Fig. 3.** Breitling Orbiter: first ever non-stop flight around the world in a balloon (1999)

from 5 to 20 years from now, based on the assumption that oil consumption will constantly rise over the years. Therefore, even a slight fluctuation in the world demand can change these figures. If we consider a middle situation, it is estimated that there are still 60 years of reserves.

However, there is an even more plausible scenario: we will never see the end of oil! The reason is very simple: because of its increasing rarity, the oil price will continuously rise until it is no longer economically viable in any way. The second reason is that greenhouse gas production will limit oil usage before its actual total rarefaction. Greenhouse gases, mostly water vapor ($H_2O$), ozone ($O_3$) and, of course, carbon dioxide ($CO_2$), contribute to global warming and are mostly emitted by burning fossil fuel.

It is estimated that, if trends are confirmed, the global temperature will increase by $1\,°C$ to $3\,°C$ by the year 2100, radically changing the global climate: ice-cap melting and thus raising the sea-level, more intense tropical storms and heat waves, ecosystem disruption, etc.

Certainly, in a hundred years time, our society will be drastically different than the one we know today. Either by force if we keep our current life-trend (in which case the outcome could be dramatic), or by choice if we decide to smooth our way out of the fossil age. This will be achieved by promoting *sustainable development*, that is

> "The development that meets the needs of the present without compromising the ability of future generations to meet their own needs. [...]"

(Brundtland Commission, 1987)

There is no doubt that science will play a crucial role throughout the transition process.

## The Solar Impulse Mission

The mission of Solar Impulse is to promote sustainable development in general, promote the use of renewable energies in particular, mobilize the enthusiasm of the public in order to change attitudes towards environmental problems, and reinforce the perception that technologies can help to achieve sustainable development. Incidentally, the goal of the solar plane project is to achieve a sustainable flight around the world with a solar powered airplane using only renewable energies (the sun in our case) to eliminate polluting emissions.

Such an airplane will have to face many challenges. The first main challenge will be to optimize energy collection and consumption. The sun radiates approximately $1.375 \, kW/m^2$, which is often referred to as the solar constant, on a surface normal to the sun. After atmospheric absorption and cloud reflection, the widely accepted solar power value that hits the ground is about $1.020 \, kW/m^2$ at sea-level. As a comparison, the sun thus radiates every minute far more energy than one year of earth power consumption. Solar power could therefore meet our energy demand if we could harness it efficiently. But the sun is, by nature, an intermittent source of energy. One key design aspect of the solar plane will therefore be to maximize energy collection by using a very large wingspan. There are obvious reasons for that: large battery storage capacity, large solar cells surface and better repartition, adequate night altitude. Two options are available: either to store the collected energy in the batteries, thus increasing the energy reserve available when we most need it (for example during the night), or to send the collected energy to the engines, therefore increasing the potential energy of the plane.

As already pointed out, a very large wingspan is necessary to maximize the energy stored during the day. The downside of this is the structural challenge. Staying as high as possible during the day is crucial for the success of the mission. Therefore, Solar Impulse must be as light as possible.

Another key problem is the wide range of temperature changes. For instance, it is estimated that temperature variations under and above the wing could reach as much as $60 \, °C$ during the day. Structural properties must remain substantially unaffected by these variations. Furthermore, batteries are not effective below a certain temperature (typically $0 \, °C$), and so temperature will have to be maintained in a narrow range using a multi-functional battery insulation inside the wing, where the batteries are to be positioned.

The last, but not the least, challenge is pilot safety. Solar Impulse will probably reach altitudes above 10,000 meters, where the temperature is as low as −55 °C, making efficient insulation and air conditioning mandatory. Furthermore, as the air gets more rarefied, the pilot must be protected from harmful radiations such as, for example, gamma rays, and a light pressurization system with low energy consumption must be inserted in the on-board devices check list. Finally, for long distance flights, an automatic control management of the airplane will have to be carefully designed to relieve the pilot from managing tedious tasks.

Overall, all these challenges make the conception of the solar plane a technological feat. Careful design decisions are crucial, and each one of them will have a huge impact on the global behaviour of the airplane.

## Mathematics for Solar Impulse

The mission envisaged for the Solar Impulse project has prompted the development of a completely new airplane concept that cannot be extrapolated from existing configurations and known design charts. Nevertheless, the advances in aircraft design achieved during the past decades and the state-of-the-art technology in the different domains have to be exploited (at least as a possible source of inspiration) to make the project realizable in a few year timeframe. In this respect, mathematical models for aerodynamics, structural behavior, energy and power management are developed and integrated in a global optimization framework. Specialists in each discipline can provide the best design in their own field, but this will not necessarily result in an optimal airplane.

Indeed, as the overall design complexity increases, local design decisions will strongly influence design in the other fields of competence. A global optimization strategy (that accounts for the different design aspects and their interactions) is the only approach that can lead to the definition of a global optimal configuration, as discussed in the next section.

In order to reduce the computational cost associated with the solution of complex physical models, special *surrogates* can be used. Surrogates are simpler models that can be applied wherever a full model would not necessarily improve the overall global model. The aerodynamic design is a typical example in which models with different levels of completeness, accuracy and computational complexity can be exploited effectively.

The design of the wing and stabilizer airfoils is driven by different considerations related to drag reduction, efficiency optimization and global airplane stability. A computationally effective model, based on a coupled potential flow-boundary layer scheme [4], has been adopted for the prediction of the two-dimensional aerodynamics characteristics (drag, lift and moment coefficients) of the wing, as well as for the laminar-to-turbulent flow transition. A more complete aerodynamic model, based on the solution of the Reynolds Averaged Navier-Stokes (RANS) equations [5], is used for the simulation of the complex three-dimensional flow around the airplane. A three dimensional CAD (*Computer Aided Design*) model is used to define the domain where we want to simulate the airflow (either the full airplane or just a portion of it). A discretization of the governing equations is obtained by subdividing the three dimensional domain around the airplane into small elements (that compose the so-called computational grid) and requiring the equations to be satisfied locally on each element (see Fig. 4). In order to capture the behavior of the flow in detail, the computational grids usually contain a large number of elements (up to a several millions), thus involving the solution of huge algebraic problems which can be tackled only by advanced parallel algorithms.

This kind of simulation is essential to achieve a good understanding of the three-dimensional effects that have an influence on the global performances of the wings,

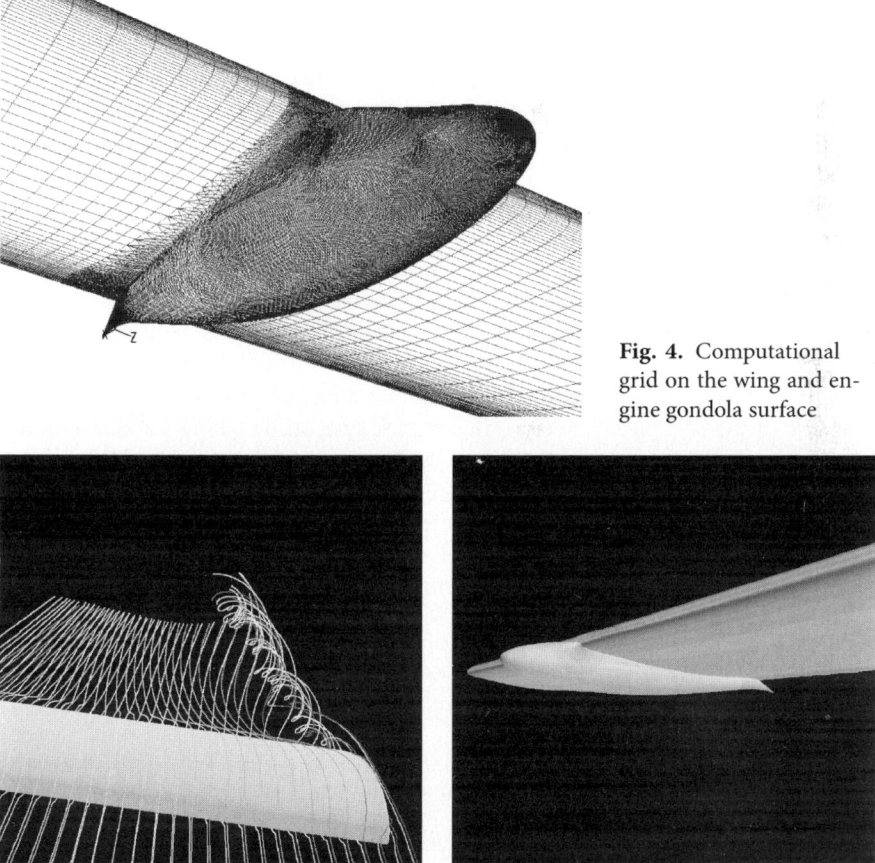

**Fig. 4.** Computational grid on the wing and engine gondola surface

**Fig. 5.** Vortex at the Solar Impulse wing tip (*left*) and pressure distribution over the wing (see the section in colour) and engine gondola surface (*right*)

nacelle and engine gondolas. Examples of three-dimensional simulations are given in Fig. 5, where the flow recirculation at the wing tip is displayed using a streamline visualization (left), while a contour plot shows the pressure distribution on the wing and engine gondola (right).

## Multi-Disciplinary Design Optimization (MDO)

In this kind of project, the design challenges are manifold and lead to difficult trade-offs. Here is an example: "How many batteries must be incorporated in the airplane?". Of course, from an energetic point-of-view, one could answer "as many as possible". There are obvious reasons for that: since solar power is the only source of energy, storing as much as possible seems the right choice to achieve success in the mission. However, batteries are heavy and an inconsiderate amount could impair the mission for several reasons: a heavy plane requires more energy to gain altitude or even sustain itself at a constant altitude; moreover, batteries must be kept at a constant temperature in order to work properly. Thus, increasing the number of batteries will result in an increase of the energy dedicated to heating purposes. Furthermore, an increased battery charge requires a stiffer airplane structure which will result in an increase of weight as well, not to mention the bending issues if the batteries are placed on the wings. In summary, determining the correct amount of battery is not easily addressed and an optimization process in several fields of expertise is necessary. Such problems can be addressed using an effective and efficient coordination of disciplinary activities, modification of conventional optimization methods, and usage of approximation techniques in a general framework called Multi-disciplinary Design Optimization (or MDO).

From a mathematical point of view, MDO is a formal process which allows the selection of the best possible design in complex systems by taking into account the interactions between various disciplines. MDO explores as many combinations as possible at every stage of the design process. In other words, MDO has to face the following dilemma: "How to decide what to change, and to what extent to change it, when everything influences everything else?"

In terms of mathematical problem formulation, MDO problems can be described using the following expression:

*we want to maximize a functional $O = M(I)$, I being the vector of design variables or inputs, M the model, O the output, under the constraint $g(I) \leq 0$.*

This is a standard description; minimization problems and equality constraints can be obtained by simply multiplying the functional by $-1$ and setting two inequality constraints respectively.

The design variables, or inputs, are numerical values that describe an acceptable design. They are either continuous (the wingspan of the plane, the angle of attack of the wing …) or discrete (the number of solar cells, the amount of battery …) and most of the time bounded.

The objectives are also numerical values that are to be either minimized or maximized. It is possible to consider multiple objectives as well. In this case, the objectives can be weighted to retrieve a mono-objective formulation. Another popular approach is Goal Programming, where each objective is maximized or minimized separately while the others are considered as constraints and kept in a certain range. This leads to computing a set of optimal, non-dominated designs represented as a Pareto curve. In this case, choosing a particular design in the Pareto curve amounts to setting a weight to the objectives. Of course, computing a Pareto curve is computationally very expensive since all concurrent designs must be computed in order to obtain the best possible approximation of the curve.

Constraints are conditions that come from either physical laws or practical considerations. A design is considered feasible if and only if all constraints are met, and therefore a careful attention must be paid to the choice of constraints.

Finally, the mathematical models are the representations chosen to relate the design variables to the objectives and constraints. They provide a representation of a phenomenon and can take various forms with respect to their complexity, ranging from algebraic to partial differential equations (linear or nonlinear). Of course, as the complexity of the model increases, the computational effort needed to obtain a solution increases too, and so a careful choice of model is often imperative.

The resolution of MDO problems is mostly based on classical optimization techniques, which include gradient-based algorithms (such as Newton's method [6]) or population-based algorithms (like genetic algorithms [7]).

Newton's methods are very popular in optimization because of their fast (quadratic) convergence rate. There might however be some pitfalls. First, the method may not converge at all. Indeed, Newton's method requires the evaluation of the functional derivative at each step, which may not be available (due to lack of regularity, or even the impossibility to compute it). In this case, quasi-Newton methods can be applied by computing an approximate derivative (using for instance a finite difference or a finite element technique). Moreover, convergence toward a global optimum is uncertain, because convergence is highly dependent on the starting point in the design space.

A radical alternative for solving optimization problems is given by genetic algorithms (GA), a special instance of evolution algorithms. These algorithms combine evolutionary concepts, such as natural selection inheritance, mutation and crossover, into optimization problems solving strategies that mimic natural evolution. In a nutshell, each parameter to be optimized is considered as a chromosome. At each successive generation, these chromosomes are evaluated according to their fitness, and then a whole new set of chromosomes is regenerated using selection, crossover and mutation, until a convergence criterion is met. The main advantages of this algorithm are manifold. The only prerequisite is a random population generation. Furthermore, no derivative or continuity assumptions are needed. Finally, because of mutations, GA are more likely to converge to the global optimum. The caveats of these methods are that they converge very slowly, mainly because the same sample may be evaluated several times and the optimum found is an approximation.

Another classical tool used in MDO is sensitivity analysis. In general, the properties of manufactured components can be described using standard probability tools. Such properties are defined using the mean value and standard deviation. MDO provides mathematical tools that sample the uncertainty interval and evaluate the objectives at the sampling points. Then, if all constraints are satisfied, designs are considered as "robust" in the sense that they are insensitive to variations in the manufactured property.

As mentioned before, in order to reduce the heavy computational time often required to solve complex physical systems, surrogate models can be used. Another way to reduce heavy computation times is to use the Response Surface Method (RMS). This is a mathematical and statistical method that interpolates (otherwise expensive) predictions within the observed space. That is, given a set of previously computed points, RSM will provide a smooth surface connecting those points. These methods provide very fast ways to compute designs; however, the approximation error generated is not always under control.

## An Example of Multi-Objective Optimization

We now give an example that demonstrates the capability of MDO to determine multiple designs under constraints. The considered problem is: "What is the maximum weight allowed for the structure with the requirement that we want to fly above a given altitude at night?". In other words, for a given minimum night altitude, what is the maximum airplane weight allowed? The purpose is to help the design team to decide, for example, what amount of glue is allowed in the solar airplane. By increasing this amount, the overall stiffness implications in terms of energy consumption are beyond the reach of the structure team. The model used was developed internally at EPFL and takes into account structural, aerodynamic and electrical properties of the solar plane. The objective, the design variables and constraints are defined in the following table.

| |
|---|
| **Objective:** <br> – Maximize the Minimum Night altitude <br> – Maximize the Structure weight |
| **Inputs:** <br> – Wingspan = 80 m <br> – Wing Area = 230 m$^2$ <br> – Battery Mass = 450 kg <br> – Structure Weight between 800 and 1,000 kg |
| **Constraints:** <br> – Minimum Night altitude > 3,000 m |

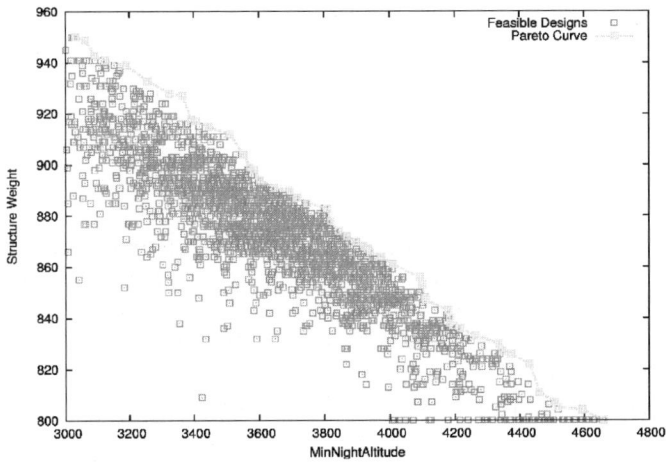

**Fig. 6.** Pareto curve for the example of multi-objective optimization

Figure 6 shows all the feasible designs computed using the solar-impulse model. There were approximately 20,000 designs evaluated and 4,000 satisfied the constraints. The right boundary of the point cloud defines the so-called Pareto curve.

As previously pointed out, selecting a point on the Pareto curve corresponds to selecting a particular design, and therefore setting a weight to the objectives. In our particular case, the Pareto curve shows the influence of weight with respect to

43

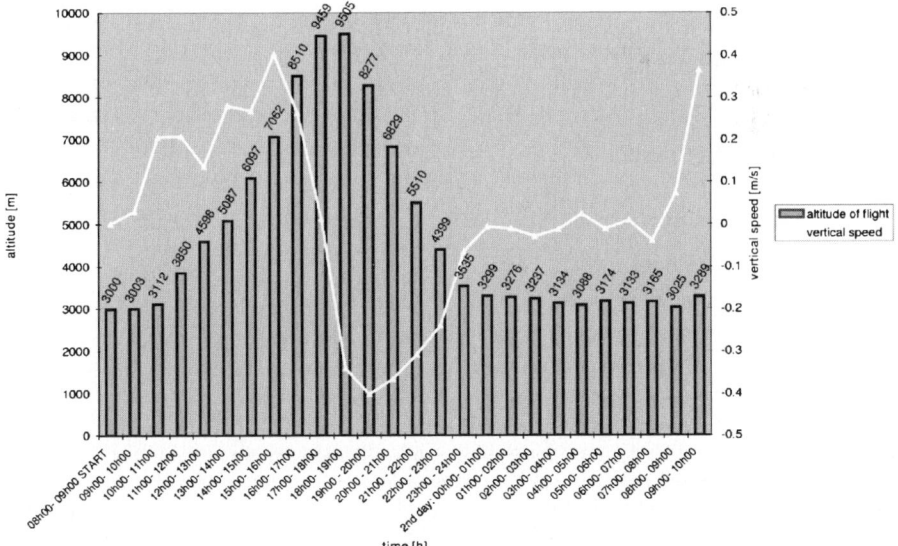

**Fig. 7.** Flight profile and vertical speed for the maximal structure weight

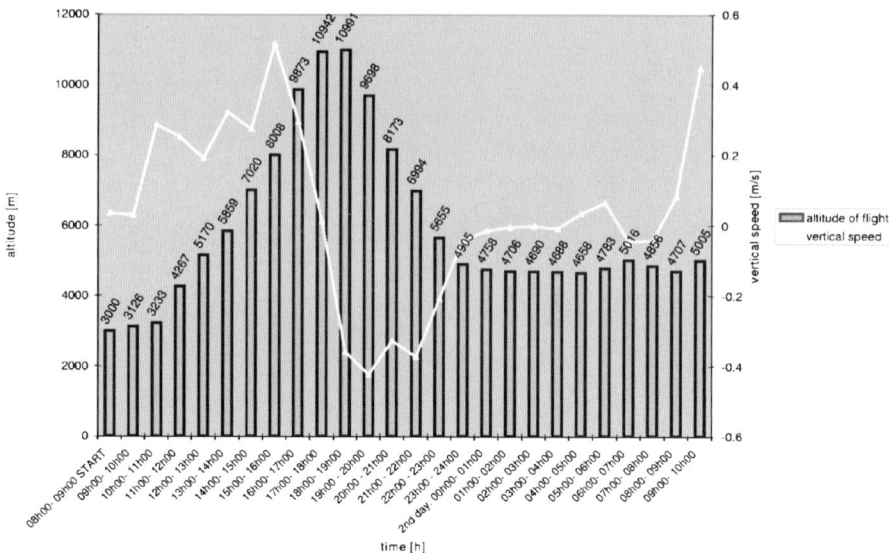

**Fig. 8.** Flight profile and vertical speed for the minimal structure weight

the minimum attainable night altitude. Clearly we can see that a structure weight of over 1,000 kg will not satisfy our constraint, and therefore the structure has to be less than 1,000 kg. If it is decided that, for security reasons, it is preferable to fly above 4,000 m at night, then the structure weight will have to be under 860 kg approximately. Figures 7 and 8 show the flight profile (altitude and vertical speed) for the two extrema of the curve. This is a typical example of how MDO can be used in terms of design decision.

## Acknowledgements

This work was supported by an EPFL grant in the framework of the Solar Impulse-EPFL partnership.

## Conclusion

Solar Impulse is an ambitious project which aims to contribute, by means of the use of advanced technologies, to the transition towards an efficient and large-scale adoption of renewable energies. In this paper, we have presented the objectives of the project and the research efforts that are carried on in the EPFL laboratories in order to sustain this technological challenge. We have analysed the role of mathematical modelling in the design process, with a particular emphasis on the optimization strategies adopted.

# Bibliography

[1]  http://www.stevefossett.com

[2]  http://solar-impulse.com

[3]  http://www.bertrandpiccard.com

[4]  M. Drela (1989) *Xfoil, an analysis and design system for low Reynolds number airfoils*, MIT Department of Aeronautics and Astronautics, Lecture Notes I Engineering 54, Notre Dame, Springer-Verlag

[5]  R. Peyret (editor) (1996) *Handbook of Computational Fluid Mechanics*, Academic Press

[6]  A.S. Householder (1953) *Principles of Numerical Analysis*, New York: McGraw-Hill, pp. 135–138

[7]  K. Miettinen, P. Neittaanmaki, M.M. Mäkelä, J. Périaux (1999) *Evolutionary Algorithms in Engineering and Computer Science: Recent Advances in Genetic Algorithms, Evolution Strategies, Evolutionary Programming, Genetic Programming and Industrial Applications*, Wiley.

45

# The Marriage Game*

Marco Li Calzi and M. Cristina Molinari

> *The happiness of a married man depends on the people he has not married.*
>
> Oscar Wilde

The marriage game is one of the world's oldest games. One takes some men and women (preferably unmarried) and lets them freely interact, hoping that they will manage to find a partner and enter into a happy and lasting marriage.

The marriage game happens to be one of the most difficult games to play, because nobody knows its rules exactly. In spite of a training phase during adolescence, the players need many years of experience and an enourmous emotional investment before they can learn the fundamental rules of the game. The game can take a cruel twist, especially for those who do not reach the final goal and end up as "bachelors". Moreover, it is only at the end of their lives that the players who get married and leave the game find out whether they have won or not. Sometimes, chance or necessity may put them back into the game against their will.

The same reasons that make the game so difficult to play probably explain the fascination it exerts on those looking at it from the outside. Obviously, there are different ways of playing the role of the observer. Readers of illustrated magazines or fans of soap opera look at this game very differently from writers, artists or movie directors, who find in the marriage game one of their main sources of inspiration.

The eye of the mathematician is particularly sensitive towards formal structures. Thus, when mathematics observes the marriage game, it is more apt to remark the characteristics that can be formalised in a model. For example, Gottman et al. [1] show how to construct dynamical models to describe and predict the evolution of a conversation between husband and wife. And Bearman et al. [2] use a graph to describe the emotional and sexual relations between about 800 students at a secondary school in a small town in the American Midwest.

---

* *Translated by Sarah Wolf*

Let us try to look at the marriage game through the eyes of a mathematician: we will construct a model (or better, a stylised version) of the marriage game and see which properties emerge.

## The Marriage Game

There are $n$ men and $m$ women. All of them have a preference ranking over the partners of the other sex, based on which they are never perfectly indifferent between two suitors. These people try to find their match within the group, or in a more romantic language, to get married. Only strictly monogamous and heterosexual matchings are admitted.

Let us consider an example with three men and four women. Brian, Charles and David are the three men. Ann, Emma, Ingrid and Olivia are the four women. For brevity, we often use simply the initial letters of the names to denote who we are talking about. Note that the initials of the men are the first three consonants ($B$, $C$, $D$) and those of the women are the first four vowels ($A$, $E$, $I$, $O$).

Suppose that Brian's preference ranking has Ann before Emma before Ingrid before Olivia. Brian would prefer to marry Ann, but he is also willing to take either Emma or Ingrid as a wife. However, rather than marrying Olivia, he would prefer to remain single. To represent this preference ranking, we use the notation:

Brian:    $A > E > I > *$

where ">" indicates preference while the star "$*$" denotes that from this position onwards Brian prefers not to get married.

For our example, let us suppose that the preferences of the seven players are the following:

Brian:    $A > E > I > *$
Charles: $I > E > A > *$
David:    $A > I > E > O$
Ann:    $C > D > B$
Emma:    $B > C > D$
Ingrid:    $D > C > B$
Olivia:    $D > B > C$

We are interested in describing which matchings can be expected to emerge. Ann, for example, is the first choice for both Brian and David. We would like to know who will manage to marry her, assuming that Ann does not succeed instead in marrying Charles, her own first choice. Of course, in general the answer depends on the circumstances. In a society where older men prevail, Ann would probably end up as the wife of whoever is older between Brian and David. In a society where women prevail, Ann would perhaps manage to take Charles for a husband.

To be more specific, let us analyse a modern (and Western) version of the marriage game where two *fundamental* rights have to be respected. The first one is that everybody can elect to remain single: nobody can impose marriage upon anybody else. Therefore, if Charles does not want to marry Olivia (as his preferences state), we are sure that the match between Charles and Olivia cannot take place. The second right is that a married person can file for divorce should (s)he find a partner whom (s)he likes better than the current one and who is willing to marry him (or her). For example, if Ann is married to David, she can divorce him and get married with Charles if the latter one wants to wed her. Of course, if Charles is already legally committed to Ingrid (whom he prefers to Ann), Charles himself is not available. Therefore, Ann stays with David only if Charles is already married off to Ingrid (or Emma).

Stable matchings are those pairings of people which respect these two fundamental rights of the individuals. In our example there are only two stable matchings:

1.  $B–E, C–I, D–A, O$ single
2.  $B–E, C–A, D–I, O$ single

Therefore, for this particular version of the marriage game, we expect one of these two configurations to emerge at the end.

In general, depending on the number of people involved and on their preference rankings, there can be few or many stable matchings. Our example had two; however, even situations with a very small number of people can have a rather large number of stable matchings. Example 2.17 in [3] reports a situation devised by Knuth where a marriage game with four men and four women admits ten different stable matchings.

Mathematicians have found several properties that hold in general for our version of the marriage game. The monograph by Roth and Sotomayor [3] collects many of these results and provides proofs. Here we choose to quote only four of these.

The first result ensures that there always exists a stable matching, whatever the number of players and their preference rankings. In other words, any marriage game has at least one solution. Stable matchings can be determined by examining the possible configurations one by one to check whether they respect the two fundamental rights. This can be very time-consuming. In the next section we describe a constructive algorithm by Gale and Shapley [4] that finds one stable matching in a very simple and direct way. The existence of a stable matching is thus a corollary of the proposition that the Gale-Shapley algorithm always leads to a stable matching.

The second theorem states that the stable matching need not be unique, as our example also shows. In very many cases all we can do is to restrict the outcome of the marriage game to the subset of configurations that are stable matchings, but we are not able to foresee exactly which one is going to emerge. Luckily, not everything is predetermined. In our example, we expect Charles to marry Ann or Ingrid but the final outcome of the game is left to chance and to the ability of the players.

The third result says that a player who happens to stay single in a stable matching must remain single in any other stable matching. In our example, Olivia stays single in any stable matching. This theorem leaves no margin for regret: those who find themselves "bachelors" in a stable matching would have remained single even if things had gone differently. Perhaps this can be a source of consolation: after all, no matter how things would have gone, Olivia could not have succeeded in finding a husband as part of a stable matching. Thus she has nothing to reproach herself about when she remains single.

The fourth theorem requires some preparation. Let us go back to our example, where there are just two stable matchings: [B–E, C–I, D–A, O single] and [B–E, C–A, D–I, O single]. We call the former M and the latter F. Let us find out how each of the men judges these two matchings.

Brian is indifferent, because he marries the same woman in both cases. Charles prefers M to F, because in M he marries Ingrid who is his first choice. And David prefers M to F as well, because in M he marries Ann who is his first choice. Therefore, for each of the men, the matching M is better than (or at least indifferent to) the matching F. The men unanimously prefer M to F.

The contrary happens for the women. Emma and Olivia are indifferent between M and F: the former marries the same man and the latter does not marry anyone anyway. Ann prefers F to M, because she marries Charles who is her first choice, and Ingrid prefers F to M, because she marries David who is her first choice. In this case the women unanimously prefer F to M. (The initials F and M help to remember which of the two sexes prefers the corresponding matching.)

Looking at the choice between M and F from the point of view of the two sexes, there is an obvious conflict. The men collectively prefer M and, similarly, the women collectively prefer F. Given that we started out from an example, this divergence in the collective preferences might just be coincidence. However, the fourth theorem states that this is an inevitable characteristic of the marriage game. If one of the sexes collectively prefers a certain stable matching to another one, then the opposite sex has an exactly contrary preference. In this case, the adage that men and women are in perpetual conflict seems to have some foundation.

## How to Find a Stable Matching

There are many ways of finding a stable matching. The most natural one, and the first that was proposed, is the algorithm of Gale and Shapley [4]. Its main characteristic is to assign different tasks to the two sexes. The representatives of one sex have the burden to propose matrimony and the representatives of the other sex have the honour to accept it. For convenience in the following description, suppose it is the men who propose and the women who accept. Later on, we will see what happens when the roles are reversed.

The algorithm proceeds in stages. At the first stage, each man asks for the hand of his first choice (if he has one). Every woman judges the proposals she may have received and chooses whether to accept one of them or to remain single. If she accepts, she will become "engaged" to the proposing man; however, the engagement is not definitive and can be broken in the subsequent stages.

At the next stage, every man who is not engaged asks for the hand of his next best choice (if he has one). Every woman chooses between the proposals she may have received, her current fiancé and staying single. If she accepts a new proposal, she breaks the previous engagement and announces a new one.

The algorithm is repeated until it reaches a state where all men are engaged or the men who are still single have asked for the hand of all women they are willing to marry. Since the maximum number of women that a man can propose to is $m$, the algorithm must terminate after at most $m$ iterations. At this point all engagements are confirmed and the weddings will (possibly) be celebrated.

We can test the functioning of this algorithm in our example. At the first stage, Brian and David both propose to Ann (who prefers David and thus becomes engaged to him) while Charles and Ingrid get engaged. At the second stage Brian – who is the only man not yet engaged – proposes to Emma, who accepts. All men are engaged and the algorithm terminates, producing configuration $M$ as a stable matching.

Now, let us switch roles and have the women propose while the men accept. At the first stage, Ann becomes engaged to Charles, Emma to Brian and Ingrid to David (who declines Olivia's proposal). In the two following stages, Olivia approaches first Brian and then David, but is always rejected. At this point Olivia has proposed to all men she is willing to marry, and the algorithm terminates, producing the configuration $F$ that is again a stable assigment.

Recall that men and women have opposite collective preferences over the stable matchings. The men prefer $M$ to $F$ and the women vice versa. In our example, the version of the algorithm where the men propose leads to the stable matching $M$; symmetrically, the version where the women propose leads to the stable matching $F$. This holds in general: among all stable matchings, this algorithm always finds the one that is collectively preferred by the sex to which the task of proposing is entrusted. Of course, more complex algorithms that are capable of finding less extreme stable matchings exist.

It is worth noting that the right to accept or decline a marriage offer is in fact less advantageous than having to make the offer. The reason is intuitively simple. The one who makes the offer starts out from his first choice and proceeds downwards, worsening his situation only when forced to. His role is active. The one who accepts offers, on the other hand, has access only to the best partner knocking on his door, but is not allowed to actively begin courting another partner that he considers to be better. Thus, he has a passive role.

Let us make two observations. The first one is that if the two sexes were to debate which version of the algorithm is preferable, the men would defend the first and the women the second one. The conflict between the sexes would shift from collective preferences over stable matchings to the choice of the method used to find one.

The second remark is of a more speculative nature. The version of the algorithm where the men propose roughly recalls the prevailing way the marriage game used to be organised in the Western world of the nineteenth century. Considering that this way favours the men, might we say that part of the progress towards equality between the sexes has been to teach the women of our century not to leave all the initiative to men?

# A Pinch of Poligamy

One of the crucial assumptions of the marriage game is that the matchings are monogamous. However, thinking of a version of the marriage game with a more exotic flavour, we could imagine that each man might be allowed to marry up to four women. What happens if we modify the marriage game allowing one of the sexes to practice poligamy?

Things become quite a bit more interesting if we change subject and, instead of men and women, speak about firms and workers or about universities and students: usually a firm hires more than one worker, while each worker is an employee of only one firm; similarly, a university admits various students, while each student is enrolled at only one university. Thus we can view their relationships as a form of marriage game in which one of the sides has the right to practice poligamy while the other one does not.

Consider $n$ universities and $m$ students. Each university can accept the enrollment of several students, possibly up to the number of positions that are available. Each student can enrol at only one university. Universities need not accept all applications they receive and students are not obliged to enrol at a university. Every student has a preference ranking over universities, based on which he is never indifferent between two universities.

Universities have a preference ranking over students with no ties, that satisfies also another assumption: the acceptance of a student by the university does not depend on who has already been admitted. In formal terms, if a university considers student $x$ to be better than student $y$ in a direct comparison, this remains true when the university has already accepted the enrollment of some students and has to evaluate whether it now prefers taking $x$ or $y$. This assumption excludes cases of affirmative action, where a student who is individually preferred ends up being rejected because the school prefers to admit a student who is ranked worse but represents an ethnic group or some other socially disadvantaged group. Our assumption imposes that the evaluation of a student depends only on his intrinsic merits and does not take into consideration also how many students of the different ethnic groups have already been admitted.

Students and universities try to combine matchings amongst themselves, or in a more bureaucratic language, to form the classes for the coming school year. This is a form of marriage game where unilateral poligamy is allowed for the universities. By means of an appropriate transformation, this problem can be reduced to a marriage game without poligamy. Therefore, many of the mathematical properties we have seen persist in this model.

In particular, the following four theorems, which are analogous to those presented for the marriage game in the monogamous regime, hold:

- there always exists a stable matching;
- the stable matching need not be unique;
- if a student is not admitted to any university in some stable matching, he will not be admitted to any university in any stable matching; furthermore, the number of

available positions that a university succeeds in filling is the same in any stable matching and, if the number of students is less than the number of available positions, even the set of students that are accepted will be the same;
– universities and students have opposite collective preferences over the stable matchings.

The algorithm of Gale and Shapley for finding a stable matching generalises in a natural way. As before, we need to assign to one side the task of proposing and to the other one that of accepting. The side which proposes is favoured in the sense that the algorithm generates its collectively preferred stable matching. Given that students and colleges have opposite collective preferences, this again poses the problem of choosing which side to entrust with the task of proposing. In this case, however, the asymmetry between the poligamous side and the monogamous side suggests to favour the latter, which appears weaker. In fact, traditionally, it is the students who apply at a university or the workers who search for employment.

## It is Not Just a Game

As in many other countries, US medical graduates have to do an internship (or residency) with a hospital department. In the first decades of the twentieth century, the competition between hospitals for the best interns and the competition between graduates for the best internships had produced a climate in which the offers of internships were made too early. In the fourties, offers were often made at the beginning of the third year of study, when there was not yet enough information on the abilities and the preparation of the student. On their part, students had to accept or reject an offer without knowing which other offers they could have received subsequently. All in all, the selection process for interns was chaotic.

Between 1945 and 1951 a serious attempt was made to establish a unique deadline for accepting an offer, but the effort did not achieve lasting effects. The hospitals continued to impose very short deadlines constraining students to decide in the dark. Hospitals, on their part, had to embark on complicated and troubling investigations because, when a student declined an offer, it was often too late to make an offer to the second choice. In 1951 the chaos was tamed by adopting a centralised algorithm for appointments, known as the National Resident Matching Program (NRMP, in the following), as reported in Roth [5].

The NRMP, which every year establishes the appointments of residents to hospitals for the whole of the United States, is obviously of great importance. However, although the algorithm worked rather well, nobody had a clear explanation of its success.

In 1962 Gale and Shapley [4], without knowing the NRMP, described the marriage game in the *American Mathematical Monthly*. In 1984, Alvin Roth [6] became aware that the NRMP was substantially equivalent to the algorithm of Gale and Shapley and rapidly understood the reason for its success: the problem of assigning residents to hospitals is a version of the marriage game with unilateral poligamy.

The centralised algorithm of the NRMP works very well because it produces stable matchings.

This intuition has allowed to update the NRMP so as to take into account changes in the characteristics of the participants. For example, one of the biggest problems recently faced is that the number of couples of residents who look for two positions in the same geographical area has increased in the course of time (cf. [7]). This is a natural consequence of the evolution of customs. While in the fifties the typical resident was single or married to a partner who was prepared to follow him, now it is rather customary that medical students meet their partner while studying. Thus couples are formed, who at the end of their studies wish to live together and therefore look for two resident positions at hospitals in the same vicinity.

Applications of the marriage game do not end here. All over the world other centralised algorithms, similar to the NRMP, have been proposed or are effectively in use, often inspired by the NRMP or the theoretical study of the marriage game and the algorithm of Gale and Shapley [4]. Among the applications in which these algorithms have been successful are: mechanisms analogous to the NRMP in Canada and Great Britain; general rules for entering the legal profession in some provinces in Canada; the National Panhellenic Conference in the US for assigning female students to social organisations known as *sororities*, including housing assignment; the market for the recruitment of football players in American university colleges.

Amongst the mechanisms recommended to improve upon the existing situation, we mention the proposal to modify the system of assigning students to universities in Turkey, that already takes place in a centralised way on a national basis [8]. Another recent and important proposal, however based on techniques that were created to solve assignment problems different from the marriage game, suggests the establishment of a centralised market for kidney exchange between patients on the waiting list for a transplant and donors [9]. This proposal has been approved in September 2004 by the Renal Transplant Oversight Committee of New England; it will most likely be possible to judge its effectiveness soon.

## Bibliography

[1] J.M. Gottman, J.D. Murray, C. Swanson, R. Tyson and K.R. Swanson (2003) *The Mathematics of marriage: Dynamic nonlinear models*, The MIT Press, Cambridge (USA).

[2] P.S. Bearman, J. Moody and K. Stovel (2004) Chains of affection: The structure of adolescent romantic and sexual networks, *American Journal of Sociology* 100, pp. 44–91.

[3] A.E. Roth and M.A.O. Sotomayor (1990) *Two-sided Matching: A Study in Game-Theoretic Modeling Analysis*, Cambridge University Press, Cambridge (UK).

[4] D. Gale and L.S. Shapley (1962) College admission and the stability of marriage, *American Mathematical Monthly* 69, pp. 9–14.

[5] A.E. Roth (2003) The origins, history and design of the Resident Match, *Journal of the American Medical Association* 289, pp. 909–912.

[6] A.E. Roth (1984) The evolution of the labor market for medical interns and residents: A case study in game theory, *Journal of Political Economy* 92, pp. 991–1016.

[7] A.E. Roth (2003) The economist as engineer: Game theory, experimentation, and computation as tools for design economics, *Econometrica* 70, pp. 1341–1378.

[8] M. Balinski and T. Sönmez (1999) A tale of two mechanisms: Student placement, *Journal of Economic Theory* 84, pp. 73-94.

[9] A.E. Roth, T. Sönmez and M.U. Ünver (2004) Kidney exchange, *Quarterly Journal of Economics* 119, pp. 457–488.

# Mathematics and Psychoanalysis

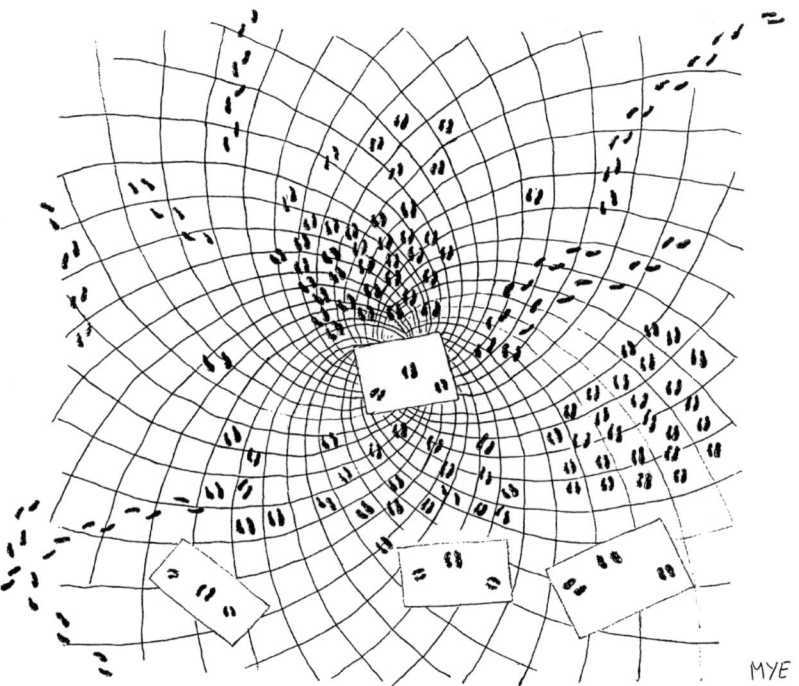

# Mathematics for Psychoanalysis.
# Brouwer's Intuitionism from Descartes to Lacan*

Antonello Sciacchitano

## No Math, No Science

It is still astonishing how much of the formalism of quantum mechanics was "already written" in the theory of the functions of complex variables a long time before the experimental data was known. Not to mention how much of Einstein's theory of relativity was pre-written in the tensor calculus of Ricci Curbastro and Levi Civita, or how much of Darwin's theory of speciation is written into the theory of chaos. It doesn't seem to be going too far out on a limb to state that mathematics is the transcendental condition of science. In other words, mathematics is the necessary condition that makes science possible. To make a slogan out of it, we might say: *no math, no science.* On this point, perhaps because he was caught up in his own cognitive fantasies, Kant was misled: mathematics is not *a priori* synthesis. It is rather the *a priori* that allows the synthesis of diverse experimental and theoretical data, because it is already predisposed to fulfil this function. Mathematics is knowledge, perhaps not yet known, as Heidegger says in *The Age of the World Picture* [1]. It is knowledge, not necessarily available in a conceptual form, that permits knowing more.

Here I would like to prove the reasonableness of this thesis by presenting how much of the logic of the unconscious is already written, certainly unwittingly, in Brouwer's intuitionistic logic.

My approach is justified *a posteriori* by the results that it produces. Thus, without justifying myself beforehand, I assume that intuitionist logic is an epistemic logic.

Traditionally the axiomatic systems of epistemic logic are constructed by addition: to the axioms of classical, propositional, or predicative logic, are added those extra that implicitly define the properties of the epistemic operator: "I know that $X$" or "I am aware that $X$", where $X$ is a propositional variable. It is therefore at first sight surprising to think of as epistemic a logic that subtracts axioms from classical logic rather than adding them to it. In fact, intuitionism suspends two fundamental

* *Translated by Kim Williams*

axioms of classical logic: the axiom of the excluded third (*A vel not A*) and the equivalence between existential and universal operators by means of a double negation: "there does not exist one that does not" is not equivalent to "all".

The first justification of intuitionism as an epistemic logic is its wealth of operators. In intuitionist logic, the three binary logical operators AND, OR, IF do not express themselves through the others by means of the negative (NOT) as in classical logic, practically becoming one only, but rather each of them maintains its own individuality that distinguishes it from the others. Further, the two quantifiers, universal and existential, are truly two and not just the same operator in different versions.

But let's look at the details. The driving idea of my work is that a classical non-intuitionist thesis, like the excluded third, becomes an epistemic operator. I will begin precisely from the …

## The Excluded Third

The excluded third is written $X$ *vel not* $X$. It is the simplest classical non-intuitionist thesis. What does it mean to say that it becomes an operator? It means that an endomorphism $\varepsilon$ is defined and that every statement $X$ is transformed in $X$ *vel not* $X$, and written $\varepsilon X$. What are the properties of $\varepsilon$? There are theorems (marked by Frege's symbol for judgment $\vdash$) and non-theorems (marked by Frege's symbol reversed $\dashv$), which we will now analyse in detail, and which characterise $\varepsilon$ as an epistemic operator. In particular, such theorems show that $\varepsilon$ shares many properties of a particular kind of knowledge: unconscious knowledge.

### Kolmogorov's Lemma

*If $X$ does not contain universal quantifiers, then $\vdash$ not not $\varepsilon X$.*

This is an important lemma, in as much as it demonstrates a kind of weak equivalence between classical logic and intuitionist logic. In fact, the declarative $X$, without its universal quantifiers, is a classic thesis if and only if its double negation is an intuitionistic thesis. In this sense, the lemma takes the coherence of classical logic back to that of intuitionistic logic, providing that the double negation is understood as "it cannot not be true that".

The epistemic interpretation is no less interesting: "it is not possible to not know that". With this the intention is not to introduce omniscience, but to state that even if you don't know now, sooner or later, given adequate analytical work, you will come to know.

Two important factors are introduced by the lemma:

– The time of knowing. We don't know all and immediately. We come to know by means of a process of elaboration that takes time.
– The epistemic value of uncertainty. Knowing is generated from not knowing. The movement is typical of Descartes's *cogito*, which acquires the certainty of the existence of the subject beginning from systematic doubt, that is, from generalised uncertainty.

For the analyst, the lemma puts in focus an essential feature of unconscious knowledge that, before being knowledge, is a knowing that is beyond the reach of consciousness, a knowledge that consciousness does not yet possess. It is knowledge that resides in the subject, where it produces effects: dreams, slips, neurotic symptoms, of which, however, the subject will become aware only when things are done, not before.

## Socrates' Theorem

Since it was dominated by Aristotelian logic, which was more interested in the transmission of the truth by means of a chain of deductions than in highlighting the knowledge embodied in it, antiquity has passed down to us very few epistemic theorems. There is practically a single one, that of Socrates, who states that he knows one thing only, which is, that he doesn't know anything. I will transcribe Socrates' theorem as a variant of Kolmogorov's lemma, where the epistemic operator $\varepsilon$ takes the place of the first negative:

$$\vdash \varepsilon \, not \, \varepsilon X.$$

For the analyst, the importance of this theorem lies in its moral connotations. The subject of the unconscious is never completely ignorant; it always knows something, for instance, it knows that it doesn't know. Faced with a completed action, the subject cannot justify himself by saying, "But I didn't know". "No," the psychoanalyst tells you, "you are always responsible, at least partially, for your actions, because you already knew something about what you did before you did it". It goes without saying that without this epistemic responsibility no psychoanalytical elaboration can begin. Psychoanalysis, before being a therapy, is an ethic: it takes care of the ethic that the subject had momentarily lost or weakened and can now, with analytical work, rediscover or reinforce.

## Descartes's Theorem

The entire movement of Cartesian doubt took place within an epistemic logic: that logic that begins with a variant of the principle of the excluded third, to be precise, the epistemic variant. In fact, Cartesian doubt can be expressed as an epistemic alternative: *I know or I don't know*, thus I am uncertain about my knowledge. From this uncertainty derives the certainty of the existence of the new subject: the modern subject of science. (I will come back to this argument below.)

It is not surprising therefore that within intuitionistic logic, which transforms the excluded third into an epistemic operator, there is a place for a theorem that can justly be attributed to Descartes: *If I don't know, then I know*. In a formula:

$$\vdash not \, \varepsilon X \, seq \, \varepsilon X.$$

The merit of this formulation is that it clarifies the nature of the knowledge at stake. We are not dealing with a bookish knowledge, written down in some manual.

Descartes's theorem doesn't serve to pass some exam at the University. In effect, we are dealing with a subjective knowledge, one that the subject actually does not know, but which he will come to know with the work of the analysis. Descartes's theorem guarantees the passage from unconscious to conscious, without saying, fortunately or unfortunately, precisely how that passage effectively takes place: the pleasure of discovering that is left to the individual subject within his own analysis.

In passing I remark that routine clinical observation will affirm that the neurotic subject, especially of the obsessive type, will state beforehand the declarative "I don't know", when he is about to declare some truth about his own story. The analyst therefore deduces: "so he knows".

There is a second formulation – philonian – of Descartes's theory, no less interesting, in that it shows the affinity with Kolmogorov's lemma:

$$\vdash \textit{not not } \varepsilon X \textit{ vel } \varepsilon X.$$

If there were no time of knowledge, there would be a tautology: either you know or it isn't true that you don't know. But since there exists a time for learning, the theorem suggests a more complex epistemic event: either you know or it is not true that you will not know, sooner or later.

### Idempotence or the Conscience That Doesn't Add Anything

The theorem of epistemic idempotence states that knowledge of knowledge is equivalent to knowledge. In a formula:

$$\vdash \varepsilon\varepsilon X \textit{ aeq } \varepsilon X.$$

If, following a suggestion by Odifreddi, I interpret the duplication of the operator as its "conscience" (see [2]), I obtain the decadence of the first Freudian topic. In effect, the distinction conscious/unconscious ceases to be operative, because the same knowledge is at work on both the conscious and the unconscious levels. The passage from one to the other is certain. It is only a question of time, even if when it will take place cannot be predicted.

### Against Ontology, or in Favour of Laity

Scholars discuss whether psychoanalysis is a science or not. Epistemic logic makes its contribution to the discussion by indicating what psychoanalysis is not: it is not religion, if it is intuitionism. In fact, in intuitionist logic the ontological argument is not valid. This means that from the existence of the object of which you know a property it is not possible to deduce that you know the existence of an object with that property: knowing the qualities of an object, as essential as they may be, is not sufficient to determine the knowledge of the existence of the object. In short, essence does not implicate existence:

$$\dashv (\exists x)\varepsilon X(x) \textit{ seq } \varepsilon(\exists x)X(x).$$

This non-theorem is a characteristic feature of intuitionism. Brouwer admits only the construction proofs of existence, furnished with an effective algorithm for constructing the object. He excludes proofs that are purely existential, without concreteness, obtained through generalisation, denying the existence of the opposite. As in psychoanalysis, it is not sufficient to know theoretically, even from authoritative treatises, that the unconscious cannot not exist; in order to know that truly it exists, it is necessary to experience it for yourself with your own analysis.

On the other hand, this logic is not obscurantist. In fact, the inverse is valid:

$$\vdash \varepsilon(\exists x)X(x)\, seq\, (\exists x)\varepsilon X(x).$$

If you know that something with a certain property exists, then there exists something of which you know that a certain property is satisfied.

## Intransitivity

The need for personal analysis is justified by another intuitionistic non-theorem: unconscious knowledge is intransitive:

$$\dashv \varepsilon(X\, seq\, Y)\, seq\, (\varepsilon X\, seq\, \varepsilon Y).$$

From the fact of knowing in theory that a certain implication is valid, for example, that $Y$ follows from $X$, it does not automatically ensue that knowledge of X follows from knowledge of Y. Knowledge of the consequence is constructed each time *ex novo*; it is not sufficient to know the antecedent. For analysis it is not sufficient to know *a priori*, because the analysis itself is *a posteriori* knowledge: it is the knowledge of the transition from unconscious to conscious, which is not automatic. In other words, there are two forms of knowledge that do not communicate: theory and practice. One alone is not sufficient. Both are necessary.

Intransitivity implies that unconscious knowledge, even when reasonable, remains strictly subjective [3].[1] The unconscious does not follow the fashion of intersubjectivity.

## Knowledge Is Not Only Knowing

The difference between cognitive logic and epistemic logic become radicalised, and at the same time is best expressed, in two theorems that represent the paradigmatic

---

[1] *The intransitivity of unconscious knowledge recalls the intransitivity of the dominance in games with more than two players. Note that our system does not even satisfy the form of epistemic self-transitivity given by the axiom of Gödel–Löb:* $\dashv \varepsilon(X\, seq\, X)\, seq\, \varepsilon X$. *By defining as episteme the formulae X for which* $\varepsilon X\, seq\, X$, *the following interesting negative characterisation of the unconscious is obtained: from the fact that you know the epistemes, it does not follow that you know of them. In this sense, the analyst who works with unconscious knowledge works with a sort of modern and not theological* docta ignorantia. *Our epistemes, understood as units of knowledge, correspond to the notion of signifier without a significance, taken up by Jacques Lacan from the great linguist Ferdinand de Saussure of Geneva.*

nucleus of two approaches. I consider Lenzen's *G* system (*G* from *glauben*, to believe) to be emblematic of the cognitive approach [4]. This is constructed by adding three axioms and a rule of deduction to the classic propositional calculation. Among the axioms is the following:

$$\vdash_G Gp\ seq\ not\ G\ not\ p.$$

This is a strictly binary axiom that counters believing with not believing. In fact, it states the *if you believe that p, then you cannot believe that not p.*

It is easy to verify that in intuitionist epistemic logic such a theorem is not valid. Indeed, the converse is valid:

$$\vdash not\ \varepsilon\ not\ p\ seq\ \varepsilon p.$$

In a certain sense this is a fundamental theorem. It states the existence of a nucleus of ignorance within all knowledge; therefore, in particular, there is no knowledge that does not know something of the negation. In other words, if not reduced to consciousness, that is, to the adjustment of the intellect to the thing, knowledge is supported by an ethical decision: it cuts from the epistemic body something that regards the negation, which turns out thus unknowable in a complete way. Freud's essay 'Negation' (1925), in which negation does not always negate but rather facilitates the return of what had been repressed, enters into this logic.

## The Double Negation

What can be said about the transformation into an epistemic operator of another classic non-intuitionist thesis, for example, the double negation?

As a consequence of the suspension of the axiom of the excluded third, intuitionism loses the law of strong double negation, which permits the cancellation of the double negation:

$$\dashv not\ not\ X\ seq\ X,$$

while maintaining the weak law that permits the introduction of the double negative:

$$\vdash X\ seq\ not\ not\ X.$$

I call $\delta$ the operator that transforms each sentence $X$ in *not not X seq X*.

### $\delta$ as an Epistemic Operator

What properties does $\delta$ enjoy? Let's list its theorems and non-theorems.

In general it can be said that $\delta$ is an epistemic operator. There are two reasons for this: first of all, it satisfies many theorems in $\varepsilon$; in the second place, the epistemic operator $\varepsilon$ is in a certain sense implicit in $\delta$. In fact, two theorems are valid:

$$\vdash \varepsilon X\ seq\ \delta X,$$

that is, every time that $\varepsilon X$ is true, $\delta X$ is true, and

$$\vdash \delta X \, aeq \, \varepsilon \delta X.$$

This last theorem establishes that "knowledge" of $\delta$: $\delta$ is valid if and only if it is known that $\delta$ is valid. Here for the second time we encounter the "uselessness" of knowledge. The discussion becomes simpler, if we drop the requisite of knowledge.

As I said, many theorems that are valid for $\varepsilon$ are also valid for $\delta$.

## Freud's Theorem

Kolmogorov's lemma becomes Freud's theorem:

$$\vdash not \, not \, \delta X.$$

This theorem becomes psychoanalytically transparent by interpreting $\delta$ as an operator of desire. Thus Freud's thesis states the necessity of desire: we cannot not desire. This interpretation is not contradicted by the theorems that follow.

## Oedipus's Theorem

The theorem of Socrates becomes that of Oedipus:

$$\vdash \delta \, not \, \delta X.$$

Forbidding incest, the Oedipus complex establishes unconscious desire as the desire that should not be desired. But, in effect, it is desired, as the following theorem states.

## Lacan's Theorem

Descartes's theorem becomes Lacan's theorem, with which it is established that even not desiring is desiring:

$$\vdash not \, \delta X \, seq \, \delta X.$$

All of these theorems regarding negation show that, in intuitionism as in psychoanalysis, negation is weak. It does not always negate, indeed, it sometimes affirms. This phenomenon was declared to be specific to the psychic apparatus by Freud in the essay on *Negation* already mentioned.

## Idempotence Is Not Valid

In contrast to the $\varepsilon$ operator, the $\delta$ operator is not idempotent. The theorem of extension is valid:

$$\vdash \delta X \, seq \, \delta \delta X,$$

but the theorem of absorption is not:

$$\dashv \delta\delta X \, seq \, \delta X.$$

As already mentioned, Freud began to construct a psychic apparatus around the first topic, based on the tripartite of conscious, preconscious and unconscious. In the 1920s he proposed another construction, based on the second topic, constituted of the tripartite of Ego, Id, and Super-Ego, where in the Id would be at work a death instinct that leads to the eternal repetition of the identical.

Just as the law of the idempotence of the $\varepsilon$ operator causes the collapse of the first topic, so the non-idempotence of the $\delta$ operator causes the collapse the second. In fact, the succession of $\delta$ operators always produces new $\delta$ operators, to infinity: $\delta^n$ is different from $\delta^{n+1}$, and the repetition of the identical is never encountered. I consider this to be an improvement of the Freudian metapsychology, in as much as it introduces the discussion of infinity into psychoanalysis.

## More About the Double Negation

In this logic there exists a third operator $\pi$ (from *psyché*), based on the double negation. It transforms each sentence $X$ into a weak form of excluded third of the type *nonX vel non nonX*(Jankov's law). $\pi$ is an epistemic operator in the sense that it is implicit in the epistemic operator $\varepsilon$. In fact, many theorems about the operator $\delta$ are formally identical to that of the operator $\pi$ (double negation, negation that affirms, expansion, etc.), but $\delta$ is not implicit in $\pi$, nor is $\pi$ implicit in $\delta$. Therefore, it is necessary to admit the psychic phenomenon that Freud calls *Ichspaltung* (splitting of the ego). In the unconscious there exist two separate and distinct threads of knowledge, independent of one another. In Lacanian terms we could say that one converges in the subject's desire, the other in the desire of the other. The two corresponding epistemic operators represent two different ways in which the unconscious "reacts" to the infinite object.

## The Subject Is Finite

I haven't given any proofs of the theorems listed, in that we are dealing with absolutely elementary deductions that can be arrived at in various ways: either with "rule + axiom" formalism in the manner of Frege, or with the "rules only" formalism of Gentzen, Beth and Kleene.

I will use the remaining space-time to give an elementary proof of the finiteness of the subject of science, which is active in the Freudian unconscious.

The subject is finite. This can be proven in various ways.

The ontological proof: The subject's existence ends with death.

The aesthetic proof: Perception of the subject is limited by the perceived object itself, of which the subject perceives always and only one part.

The linguistic proof: The subject of the declaration is finite because every declarative act puts at stake in the "here and now" only a finite number of signifiers.

The logical proof: In my opinion this is the most convincing proof, as long as the logic is epistemic. This is more correct that the aesthetic in that it does not confuse finiteness with limitation. Spinoza had already proven that there exist infinite sets in which the elements are limited, such as the set of distances between circumferences of non-concentric circles.

The proof begins with Descartes. Cartesian doubt, stripped of the rhetorical frills that Descartes liked to dress it up with, reduces to the epistemic alternative: *either I know or I don't know*. The reasoning then continues, "If either I know or I don't know, then I am a subject who doubts." But "either I know or I don't know" is true, thus by *modus ponens* "I am a subject that doubts" is true. We ask ourselves: "When is 'either I know or I don't know" true?" Here we know to answer. "Either I know or I don't know" is an instance of the excluded third in epistemic form. Brouwer has by now incontestably proven that the excluded third *A vel not A* is valid only in finite universe. *Ergo* it is not wrong to state that the subject, who depends on an epistemic form of the excluded third, is finite. The Brouwerian example is simple. If I have two sets $A$ and $B$ and I verify that their union $A \cup B$ is formed of eleven elements, then I can affirm that either $A$ is greater than $B$ or $B$ is greater than $A$, to the exclusion of any third possibilities. That certainty would be diminished if the union had a number of elements that was even or infinite.

## The Object Is Infinite

Since the time immemorial of Greek logocentrism, logic appears to be without an object. Mathematics, on the other hand, is not without an object: infinity is the object of mathematics. Because it is more mathematics than logic, intuitionism too works on the basis of the infinite object. This can be confirmed in its semantics, which, as Gödel [5] predicted and Kripke [6] realised, avails itself of infinite ordinal models.[2]

From our point of view, it is sufficient to recognise that the relative certainty of the finiteness of the subject is the point of departure for reasonably verifying that the object to which the subject of the science relates, and with it, the subject of the unconscious, is infinite. It presents itself as the spatiotemporally infinite in physics, as biodiversity in biology, as the object of desire in psychoanalysis. But how can we conceive an infinite object? Freud tried to do it by means of the infinite repetition of the identical. But that is a poor solution. Are there others? Yes, there are infinitely many others. The problem of the infinite is largely indeterminate. Let me explain better.

The infinite is a structure of modern episteme, which Oswald Veblen in 1904 proposed calling non-categorical [7]. This means that the structure in itself cannot

---

[2] *Without going into details about Kripkean semantics, I would like to note that an intuitionist model is a countable set of epistemic states, partially ordered by a reflexive and transitive relationship. I also note that for the semantics of classical logic models with a single epistemic state are sufficient.*

be represented – Freud would say that it remains within the primal repression – but it is possible to make partial models or representations of it that are not equivalent to each other. In the (uncertain) terminology of Freud, the models of the structure would be examples of a return from the repression. The eternal repetition of the identical itself is a model of the infinite. This is a different model from numerable infinity, made up of infinite numbers that are all different, and also different from continuous infinity, made up of points so densely stippled that there are no gaps. The infinite repetition of the identical served Freud to explain the existence of an unconscious feeling of guilt, somewhat like a numerable infinity serves to count, and continuous infinity to draw and measure things on earth.

The result of non-categoricalness is interesting for several reasons. First of all, it distinguishes scientific infinity from religious infinity. In fact, religious infinity is single, as testify the great monotheistic religions. On the other hand, scientific infinity is plural, and its plurality conditions two aspects of scientific discourse: indeterminism and self-revision. Scientific indeterminism is testified to, for example, by quantum mechanics and by the function of chaos in biology. The acceptance of the indefinite revision of scientific theories is a fact in modern epistemology, from the historic epistemology of Bachelard to the falsificationist epistemology of Popper. Scientific theories are not incontrovertibly and definitively codified in some treatise, but live in the perpetual renovation of the social bond within the scientific collective. Indeterminism is valid in psychoanalysis as well, for example, in sexual relations;[3] in the same way, it would also be found in the self-corrective recovery of metapsychology in the psychoanalytic discourse, if it were truly scientific and if it lived in the collective of scientific thought. This is what I am hoping for, and I am working so that this happens in these times when it seems to be particularly 'in' to talk about the death of psychoanalysis.

The proposal of intuitionism as the mathematics of psychoanalysis exactly suits this purpose.

## Bibliography

[1]  M. Heidegger (1950) "The Age of the World Picture", in: *The Question Concerning Technology and other Essays*, William Lovitt Ed., Harper Torchbooks, San Francisco 1977, pp. 117–118.

[2]  P. Odifreddi (2003) *Il diavolo in cattedra. Logic from Aristotle to Gödel*, Einaudi, Turin, p. 21.

[3]  J. von Neumann, O. Morgenstern (1947) *Theory of Games and Economic Behavior*, 2nd ed., Princeton University Press, Princeton, pp. 38–39.

[4]  W. Lenzen (1980) *Glauben, Wissen, und Wahrscheinlichkeit. Systeme der epistemischen Logik*, Springer Verlag, Vienna, p. 142.

[3] *The sexual relation is indeterminate, that is, it admits infinite solutions, like a system of two equations in two unknowns that differ by a multiplicative constant. There are reasons that justify this as an opportune correction of Jacques Lacan, who rather states the non-existence (or impossibility) of sexual relations, but a discussion of this is beyond the scope of this present paper.*

[5]  K. Gödel (1986) *On the Intuitionistic Propositional Calculus* (1932), in: *Kurt Gödel: Collected Works. Volume 1*, edited by S. Feferman, J.W. Dawson, S.C. Kleene, G.H. Moore, R.M. Solovay, and J. van Heijenoort, Oxford University Press, Oxford, 1986, pp. 223–225.
[6]  S. Kripke (1965) Semantical analysis of intuitionistic logic, I, in: *Formal systems and recursive functions*, North Holland, Amsterdam, pp. 92–130.
[7]  O. Veblen (1904) A System of Axioms for Geometry, *Transactions of the American Mathematical Society*, 5, pp. 343–384.

# Mathematics and Applications

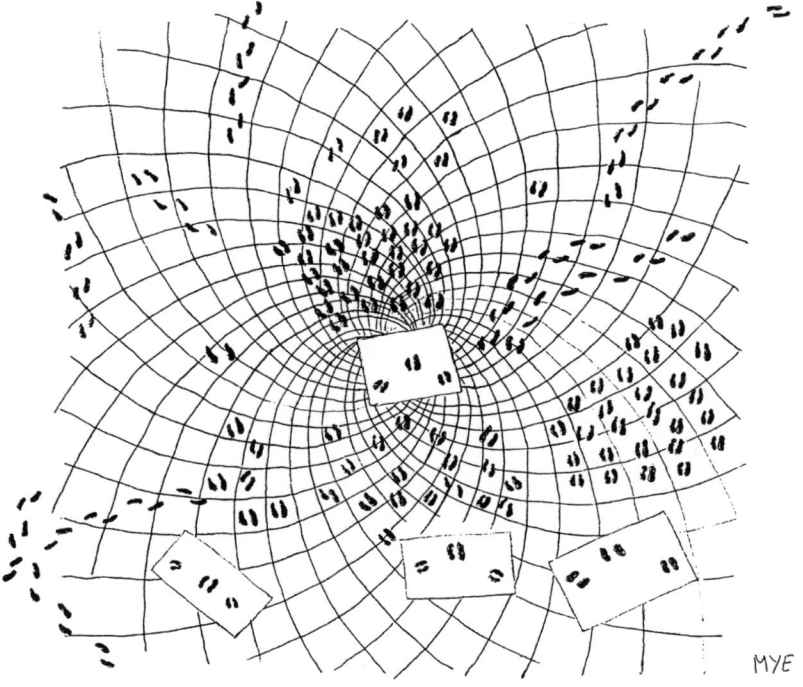

# Search Engines and Mirrors of Society*

Massimo Marchiori

## The Mirror

The web is becoming an increasingly important reality, a truly parallel virtual world and, as such, merits analysis and study. Given the mass of information present on the Web, the main tool used by everybody to access this information is the search engine, the main "television channel" that allows us to access the information present on the Internet.

Now, when seen from certain points of view the Web represents our society, and is therefore a kind of mirror for reality. It is thus natural to try to use search engines as true mirrors for society. This is, for example, what has been done for some time, analysing what questions people ask to search engines. This way we obtain a cross-section of society's tastes and interests, a kind of on-going universal poll. Google itself offers part of this information in a section called *Zeitgeist*, a German word that means, more or less, "spirit of the times". In the *Zeitgeist* it is possible to find out what is most frequently asked to Google, so as to quantitatively shed light on people's main preferences. For example, at the beginning of this year the most popular actress in Italy (on the Web) was Angelina Jolie, followed by Jennifer Lopez and, in third place, Monica Bellucci.

This kind of statistical analysis is certainly very interesting, and is veritable manna for publicists, sociologists, historians and analysts. However, it stops at one side of the story, while in fact things are much more complex and interesting.

## Only a Mirror?

The analysis of the questions asked to a search engine views the engine itself as a mirror of society and the Web as our reflection. But are we sure that the search engine

---

* *Translated by Kim Williams*

is only a mirror? Or is it rather something more, something that links two realities, making them interact with each other in a complex way? In other words, perhaps what appears to be a mirror really is not, and we can try to go through it. In order to do this it is necessary to try to consider not only our society, on one side of the mirror, but another society as well, the Web society, on the other side, and to see if the search engine acts solely as a mirror in this context, or as something more.

In this scenario there are three fundamental observations to bear in mind:

**Fig. 1.** Alice Through the Looking Glass (John Tenniel, 1871)

*First*: our reality influences the Web.
This is an obvious point, which confirms how much the Web, in its complexity, was born from and constantly depends on the input provided to it by the external world. In other words, the other side of the mirror changes according to what we do.

*Second*: the Web itself can influence our reality.
This derives from the fact that the Web is increasingly an *information medium*, a means of information, and as such has repercussions on our world as well: sometimes what appears on the other side of the mirror changes some things on our side too.

*Third*: the search engine is not merely an observer.
The search engine is not only a mirror placed between the two worlds, but is subject to a kind of Heisenberg's indeterminacy principle: when it reflects something, it necessarily modifies the reality that it is reflecting as well (both immediately, and in the longer term).

## The Web and Society

Let us therefore go further, trying not to see the Web simply as an object, but as a true alternative world (how linked it is to our society remains to be seen).

Widening our perspectives, let us consider the Web as a true society. The first thing to decide is who the inhabitants of this society are. Obviously, depending on this choice, the model of society that we will analyse changes. One reasonable choice is that of considering the Web pages themselves as inhabitants of the Web society. Each Web page can be seen as a person, and indeed, to distinguish Web people from "real" people (those who inhabit our society), let us give them a different name, *p-erson* (where *p-* reminds us that we are talking about Web *p*ages). What is therefore the link between the two societies, the human one and that of the Web? At first glance we can immediately associate every person to his or her p-ersons: each person is responsible (wholly or in part) for the creation, content and maintenance of various p-ersons (for example, our personal Web page, or documents, parts of commercial and institutional sites, and so on).

It remains to be established what is the "primary engine" that directs the evolution of the Web Society. In other words, what is the "life goal" of the p-ersons? When can we say that a p-erson is successful? A reasonable definition of success for a p-erson is the number of people who visit her. This means, given the size of the Web, that in order to be really successful it is necessary to have a good ranking (a positive evaluation) by the search engines. In other words, at least as a first approximation, the success of a p-erson can be measured by its ranking by the search engines.

Now, by far the most used search engine in the world is Google, which uses as a measure the by-now famous PageRank [1]. And thus, again with a good approximation, we can identify the success of a p-erson with her PageRank.

## Generalised Democracy and Demokracy

What kind of society is the Web? Can we compare it to the models that have appeared in our society, such as, for example, democracy, monarchy, or oligarchy? Or is it something different?

Our present society of reference (at least as far as the majority of states in the world is concerned) is democracy, and therefore the natural question to ask is: Given that our society is prevalently democratic, is the Web too, by reflection, a democracy?

In order to answer this question we need more general definitions, and it is necessary to broaden our horizons, to go beyond the definitions of society that we are used to.

Let us take then the concept of classic democracy (a banal definition of which is "one man, one vote").

In classic democracy there are three fundamental characteristics. The first is the typical one, if you will, of democracy: *equality*. As mentioned, one man, one vote, that is, all people are equal and therefore, in terms of voting power, to each person corresponds one vote. The other two are so obvious that, instead, they are often not even taken into consideration, partly because they are integral to and rooted in the voting process (at least in a first approximation, given that, due to fussiness, some electoral laws complicate the situation): *atomicity* and *additivity*. Atomicity means that the vote is indivisible: it is impossible, say, to give 1/3 of our vote to one candidate and 2/3 of it to another. Additivity means that the votes are combined in an additive way (meaning that one vote plus one vote equals two votes … [in democracy $1 + 1 = 2!$]).

Let us now define *generalised democracy*, where these two properties are relaxed. In generalised democracy equality is still valid (which characterises a democracy, but atomicity no longer is (thus the vote can be distributed) and additivity is substituted with the weaker property of monotonicity (that is, composing $X$ votes with $Y$ votes will give a result greater that both $X$ and $Y$, but not necessarily $X + Y$).

We can also go beyond generalised democracy, and relax the first fundamental property as well, that of equality. Let us define a new class of democracies, **K-democracies** (or, more briefly, **demokracies**), where K is a number between zero and one ($0 < K \leq 1$). In these democracies the rule of equality is no longer valid, and so people can have diverse voting powers: from $K$ to 1. Thus, for example, the 1/2-Democracy (half-democracy) is "half democratic", since that some people can have a voting power (1/2) that is half of that of the others (1). And, as $K$ diminishes, we go further and further away from "true democracy" (1-Democracy).

As in the case of generalised democracy, we can also have **generalised K-democracy** (in which all the three hypotheses of equality, atomicity and additivity are relaxed).

## Democracy in the Web

Since we now have at our disposal much broader definitions of democracy, we can look again at the question of what kind of society the Web is. To what does a vote correspond in Web Society? For what we have said above regarding the goal of Web society, and thus its equivalence with a high PageRank, the correspondent to a vote is the link that a p-erson might or might not give to another p-erson. Now, there are two points of view, one internal and one external to the Web. The first has as its principal agents the p-ersons, while the second, on the other hand, considers the Web from the point of view of the persons who in turn control the p-ersons in the network.

Let us then see what happens from the internal point of view; in this scenario the following property holds true:

The Web (with Google/PageRank) is a **generalised democracy** for the p-ersons.

This is, we might say, the "formal" confirmation of the fact that Google is perceived as an impartial search engine, and thus not only offers an undistorted vision of the Web, but its intervention as a measure of success in the Web is reasonable, in the sense that from the internal view of the Web, it forms precisely a kind of society that is fully democratic.

Let's see, on the other hand, what happens when we consider the external point of view, that is, when we consider not only p-ersons, but persons as well. If the Web population does not change, we have the following result:

The Web (with Google/PageRank) is a **generalised K-democracy** for persons when the only changes are local to the p-ersons.

However, in general the population of the Web is not immutable, and so what happens when we consider the Web Society in its full generality?

The Web (with Google/PageRank) **is not a generalised K-democracy** for the persons.

That is to say, at the moment when we also take into consideration our society (the persons), the Web Society (which would be a generalised democracy) is "corrupted", losing its democratic character, and also dodging the very bland concept of generalised demokracy.

Therefore, if the Web as a whole is not a democracy, what kind of voting does take place? The point is that the voting as a whole is principally constituted of three classes of votes:

- "Normal" (true) votes; these are the votes that are effectively democratic.
- "Exchanged" (cooperative) votes; those that take place through a consortium of persons and p-ersons who vote for each other, creating genuine cartels.
- "Bought" votes, that is, votes that are acquired through monetary transactions by people in our society.

Because of its enormous size, the Web, at least for now, has the overall property:

$$(N + E + B) \gg (E + B)$$

that is, the total sum of votes (normal plus exchanged plus bought) is much greater than the sum of exchanged votes and bought votes. This is what makes the Web still a true social reality and not a fake society.

## Collateral Effects

The results of the preceding section are purely theoretical: the fact that the Web for persons is not even a generalised demokracy does not mean the Web, in practice, has not behaved in a democratic manner. In fact, because of the principle of Heisenberg that we mentioned previously, the Web was for some time a democracy, then slid progressively towards generalised demokracy, and subsequently lost even this characteristic, and recently there has been a proliferation of "artificial links" (exchanged and bought votes) that have largely defaced the Web's democratic structure. The birth of exchanged and bought votes is part of a wider activity that goes by the technical name of *Search Engine Persuasion* (SEP), that is, the group of techniques developed to manipulate the ranking of the search engines (in general, to artificially, and so undeservingly so to say, making some p-ersons more successful; see for example [2].) Today SEP is also known as SEO (*Search Engine Optimisation*), a name which is actually misleading and created more recently by the flourishing market of firms that offer this service, where it would appear that they are optimising (a politically correct term?) the results of the search engine, whereas in reality we are really dealing with manipulation.

What is certain is that, whatever name is used to indicate this activity, in today's world there is an entire segment of the market dedicated to SEP (and thus to the creation of exchanged and bought links aimed at destabilising the democratic nature of the Web Society). It is also interesting to note that, until now, this market is prevalently localised in the commercial sector (that is to say, that of sales of products and/or services), but there is nothing to prevent that in the future, with the increasing importance of the Web Society in our society, this anti-democratic trend will also touch other aspects such as information, politics, culture, religion, and so on.

"Amateurish" examples (because they are announced and therefore evidently forced) of this trend are, for example, the so-called *Google bombings*. To give some concrete (and by now) famous examples, up to a short time ago someone searching for 'miserable failure' on Google would have been directed to the Web page of the then-President of the United States, George W. Bush. The same kind of bombing also occurred in Italy, where searching for the Italian equivalent of 'miserabile fallimento' resulted in the page of the then-Presidente del Consiglio, Silvio Berlusconi. Similarly, until a short time ago, the top result of a search for the word 'devil' was Bill Gates, the founder of Microsoft.

It is evident that such examples are important for showing the potential of SEP, and they should not be brushed aside as mere high jinks: the real danger is that techniques such as these can be used in the future not only for propaganda or practical jokes, but for much more subtle manipulation of Web Society and the information it makes available to our society. At the moment when the Web imposes itself as the dominant medium in the information sector for the majority of people (a role presently occupied by the traditional media: television, print, radio), it is obvious that the risks of an anti-democratic trend in a virtual and therefore much more easily manipulated society, threaten to make the Web Society a very appetising conquest.

## Displacements

In addition to the distortions due to exchanged and bought votes, it is also necessary to point out that there are displacements from the norm that are not due to particular interests (and thus, that are created from the normal votes) but which depend on the way in which the search engine works, which, like all absolute measures of success, is necessarily fallacious in multiple situations.

There are various kinds of displacements, such as for example the types called *Yin-Yang*, *Glue*, and *Bubble*. The *Yin-Yang* typology happens when the assumption that a link connects p-ersons to other similar p-ersons collapses, and indeed, even the opposite happens: the links purposely connect p-ersons with opposing attitudes. One very famous example occurred when for a certain period in the history of Web Society almost all sites with pornographic material were linked to sites that were diametrically opposed, such as, for example, sites for cartoons (Walt Disney) and p-ersons suitable for minors in general. This happened because sites for adults asked visitors for confirmation that they were legally of age, and in the negative case they diverted (linked) to sites suitable for minors.

The other kind of displacement is the so-called *Glue*, where the connection (link) takes place for reasons that don't make common sense, but are maybe metaphorically motivated, or in any case much more general than the intended context. For example, up until a short time ago a search for images of Lord Byron in Web Society would have turned up images of an enticing young woman, because one of her admirers had dedicated several poems by Lord Byron to her.

Another example, even more common, are the so-called 'temporal bubbles', which change the web for a limited amount of time, 'peaks' of deviations from the norm, and are then re-equilibrated by the Web system. For another example involving Lord Byron, a search for information about the English poet in the period following the 2002 world soccer championships turned up pages of images that had practically nothing to do with the famous bard, being dedicated instead to a completely different Byron, Byron Moreno, the soccer referee who became famous at the time for a much-discussed (euphemism) call during a soccer game that resulted in Italy's elimination from the competition.

## Evolution

Another aspect to take into consideration is the evolution that the structure of Web Society has undergone over time. Everyone knows that the Web grows almost exponentially from year to year. However, what is often overlooked is that there is not only a growth, because like in any proper society, a part of the population disappears and dies over time. In this respect the Web differs from our society, because while in human society the standard deviation of maximum age (that is, the differences in life expectancies) is relatively low, in the case of Web Society the differences can be enormous.

In Web Society there are therefore many differences in caste, so to speak, and it is interesting to bring the analysis further, studying which individuals (p-ersons) are on the average less fortunate than the others. This point of view is common to both our society and Web Society, even if it isn't a particularly exalting one: in Web Society the p-ersons that are 'poor' in links (that is, approximately, those that are less successful) die (that is, have a much shorter life span) much more than the other ones. Thus in the Web too, sadly enough, the 'poor' life is worse and they risk dying much earlier than the 'rich'.

The analogies with our society don't stop here: in fact, besides mortality, it is interesting to see how 'wealth' (success) is distributed in Web Society. And here too we discover a situation that is very similar to our own, as there is a notable discrepancy between the rich tiers and the poor tiers, between successful p-ersons and those who are not successful. The type of discrepancy is the same in both societies: in other words, the distribution of 'wealth' tends to follow a *scale-free* course, that is, one that is invariant in scale [3]. This implies that as we climb up the wealth ladder, the number of p-ersons diminishes exponentially. Thus, the distribution of wealth is strongly unbalanced in Web Society too.

Furthermore, another interesting, correlated thing is that it appears that the Web follows another aspect of our society as well: the so-called *small world structure* [3,4], essentially a society in which, through 'friends of friends', each person/p-erson can efficiently come into contact with other persons/p-ersons (perhaps to ask a favour, as in our society, or to increase the ranking of our associates, as in Web Society). None of this is obviously random, given the driving role that Google has played in this whole evolution, contributing to the improvement of Web connectivity (for a historic retrospective, see the concept of 'arena' in [5], comparing it to what happened successively).

79

## Theology

Up to now we have seen how the role of search engines is much greater than a simple mirror of society: they constitute an essential pivot of Web Society and thus, by reflection (!), of our society. But how far can we go in the evaluation of the importance and the impact of search engines? Thomas Friedman, a well-known editorialist, fixed the limit in his famous 2003 editorial in *The New York Times*, with the so-called 'Friedman assertion', which says, and I quote, 'Google, combined with Wi-Fi, is a little bit like God'. This thesis immediately became the object of great polemics and controversies, which paradoxically (also because of the Yin-Yang effect mentioned above) only made the popularity of the thesis grow. Beyond the thesis per se, which is obviously an evocative slogan, what Friedman said in his editorial (which evidently few of the critics who labelled him an idiot had read completely) is reasonable: the importance that the search engine (whether Google or some other) will have in the future, in a future in which our connections to the net can occur everywhere, will be of enormous proportions, comparable to a modern oracle whom we can ask questions and obtain answers.

And in fact, this is more than reasonable, almost obvious, given the evolution that the Web Society is undergoing as a container and provider of information of all kinds. Few today would disagree with the fact that the most important medium, television, is also a formidable instrument for opinion and also plays an active role, in addition to a passive one, as far as society is concerned. This is even more of a reason why this should happen with the search engines of the future, since they will be the new 'television channels' of Web Society, those channels that are the mediators between users and the mass of information.

In a certain sense this is inevitable, and even more so on the Web, where the enormous quantity of information offered makes necessary the presence of an intermediary who offers a selection and a search of information at very low cost for the user. And this is not necessarily a bad thing, on the contrary! Where, then, does the danger lie?

## Democracy

As in all the media that deal with information, the real danger isn't so much in the media per se, as much as it is in the regime of monopoly. When there is only one channel for information (only one channel on "tomorrow's television", which is Web Society), it is obvious that there is an extremely high risk of manipulation of the information.

And in any case, from what we have seen up to now, it seems that after all the search engine per se is the lesser evil: PageRank makes Web Society, internally, essentially democratic, and it is then the presence of our society (persons) that 'corrupts' in some way this Eden, overthrowing the democratic rules.

This is true, except for a detail of fundamental importance: ***Google, currently, does not use PageRank at all!!***

This is due in part to a process of natural adaptation, if you will: part of the Web Society has adapted to Google in order to be successful, in many cases in an underhanded way (SEP), and thus one way out is for Google itself to evolve, in order to deal with the changes that it has provoked. And also to deal with the various displacements (see the corresponding section above) that initially were not taken into consideration.

Thus a certain degree of evolution and 'improvement' is, we might say, completely natural. What is much less natural, from the point of view of the rules of democracy, is that after the introduction of PageRank (around 1998), Google has never again revealed what algorithm is used.

To be precise, Google has used a variant of a technique initially set forth in [5], that is, a part of its technology (PageRank) is in the public domain, while another part is a jealously guarded secret. This makes SEP much more difficult, although it happens anyway (except that now, performing SEP has become much more costly and thus not everyone can afford it). But the most serious aspect is that all this raises a doubt about whether the algorithm used today by Google is truly democratic or not. In other words:

Is the Web (with Google) at least a **_generalised demokracy_** for p-ersons?

The answer, unfortunately, is quite clear: NO!

This answer is known to the professionals who study the behaviour of the 'secret rules' that Google uses to give its measure of success.

For example, it is well known that Google penalises the p-ersons that are blogs, that is, Web pages that use blog technology to style personal diaries that can be linked to the ideas of other persons in other diaries. We might say that, in a certain sense, Google is racist.

Another example is given by the fact that apparently Google, internally, seems to assign a measure of success greater than 1 (which is the maximum success that can be obtained by a p-erson if PageRank is used) to 'its' p-ersons, that is, to Web pages doing business with it. And given that Google's business model, that is, the way that Google supports itself, is through advertising, there are very strong doubts as to whether large firms that are advertising partners of Google receive similar treatments, bonuses of success that are completely anti-democratic.

Moreover, certain behaviours in the creation of links have been classified as penalising by Google: it was later discovered that these penalty rules were not applied to Google associates.

Google has its team of 'favourites'.

And dually, there are strong doubts as to whether the companies that do not collaborate with Google, or are in competition with it, are not given a penalty with respect to the others (this has already been demonstrated with regards to Google's market partners in its advertising programme for the sale of keywords).

What is disquieting about all this is that for the great majority of the public, and many of those who are well informed too, Google functions through PageRank, and thus a measure that is substantially, in itself, democratic, when instead this has not been the case for years, and decisions that are totally arbitrary have been made, in a regime that is essentially 'dictatorial', such as how to regulate the success of

81

p-ersons in Web Society. To unhinge this false impression a strong competition in the market is necessary, but the danger is amplified by Google's monopolistic position, which by now has acquired its base of users. And it is not likely that the introduction of another large search engine, which might be that of Yahoo, would solve much of the problem, since in that case as well the rules by which 'success' (ranking) is assigned are jealously guarded secrets, and decided in a dictatorial regime. Thus the choices are between one, or two or three, 'benevolent dictators'.

The huge risk is that the search engines (for now Google with a monopoly, perhaps two or three in the future) will not be mirrors at all, or impartial arbiters, but will play the role of 'modern dictators', giving the impression to the great masses that democratic rules are controlling the game, while instead they will be deciding how to manipulate information. Friedman was most likely very wrong: instead of something that is a little like God, we run the risk of finding ourselves with something that resembles very much a deceiving demon. Not a democracy, but a ***demoncracy***: apparently only different by an 'n', but in reality, a world of information that risks being distorted.

Mirror, mirror, on the wall …

## Bibliography

[1] Sergey Brin and Larry Page (1998) The anatomy of a large-scale hypertextual Web search engine, in: *Proceedings of the Seventh WWW Conference*, Brisbane, Australia.
[2] Massimo Marchiori (1997) Security of World Wide Web Search Engines, in: *Proceedings of the Third International Conference on Reliability, Quality and Safety of Software-Intensive Systems (ENCRESS '97)*, Chapman & Hall.
[3] Albert-Laszlo Barabasi (2003) *Linked: How Everything is Connected to Everything Else and What it Means for Business, Science, and Everyday Life*, New York, Plume.
[4] Mark Buchanan (2002) *Nexus: Small Worlds and the Groundbreaking Science of Networks*, New York, W.W. Norton.
[5] Massimo Marchiori (1998) Enhancing Navigation in the World Wide Web, in: *Proceedings of the 1998 ACM International Symposium on Applied Computing (SAC '98)*, World Wide Web Applications Track, ATM Press.
[6] Thomas L. Friedman (1998) Is Google God?, in: *The New York Times*, 29 June 2003.

# Mathematics and Cells: Brief Tales of Chemotaxis, Neurons and a Few Digressions*

GIOVANNI NALDI

> *"What is the role of mathematics, professor?"*
> *"Mathematics has a central role in culture ... "*
> *"That is to say ... "*
> *"... that the ideal of all sciences is to become mathematics. The ideal of the laws of physics is to become mathematical theorems, mathematics is 'the end of the science'."*
> *"Even biology?"*
> *"Yes, even biology [...] Mathematical thought is clear thought ... "*
>
> Interview with Giancarlo Rota

> *"Mathematics is not an empirical science, and yet its development is strictly tied to that of the natural sciences ... indeed, it can't even be denied that some of the best inspirations in mathematics, in those parts of it that constitute pure mathematics as one imagines it to be, come from the natural sciences ... "*
>
> J. von Neumann (1947)

The relationships between mathematics and biology have a long history, in which have participated G. Galileo, W. Harvey, R. Boyle, R. Hooke, L. Euler, T. Young, J. Poiseuille, H. von Helmholtz, D. Bernoulli, F. Gauss, N. Wiener, J. von Neumann, V. Volterra, and many others. Certainly these relationships are not, for various reasons, always easy, even if analysis and numerical simulations of mathematical models in the more general context of life sciences is slowly proving itself to be an ulterior investigative instrument to be used in conjunction with experimental or theoretical methodologies. In a mathematical model the real phenomenon that is to be investigated is represented by quantities that are typical of mathematics (functions, equations, ...) placed in relation to each other on the bases of known hypotheses and notions about the phenomenon. Among the advantages that a mathematical model in biology should have we can recall: predicting the evolution of a biological system in different conditions without repeating experiments or in situations that are not

---

* *Translated by Kim Williams*

verifiable experimentally; validating biological hypotheses quantitatively; investigating properties of biological materials. In each case there is a risk, to quote Manfred Eigen (winner of the 1967 Nobel Prize for Chemistry), "a theory has only the alternatives of being right or wrong. A model has a third possibility: it can be right but irrelevant". The diffidence regarding the relevance of a model, and thus of its capacity to predict or reproduce a phenomenon is one of the causes of difficulty in the interactions between the mathematician and the biologist (in the final analysis, biology has been developed without the consistent contribution of mathematics, in contrast to physics) and is the theme of various jokes in which the mathematician and the theoretical physicist are mocked for their solutions.

In any case, the relationship between the phenomenon (the reality) and the model certainly cannot be that of a precise description in every detail, as happens in a story by J.L. Borges in which the emperor, intent on having a precise geographical map, thinks to ask the cartographer for a meticulous description of every detail in which nothing at all is overlooked; the map would become a paper territory at full scale: completely useless for orienting oneself. Beginning with knowledge and observations that are biological, chemical, physical, etc., one of the objectives of those who develop mathematical models consists in the possibility that these might be predictive (to be able, beginning with the observation of the real phenomenon, to make predictions or deductions, or to reinforce theoretical hypotheses; to quote Anaxagoras "the vision of things that are hidden is given by things that are manifest"). Sometimes what are sought are "postvisions", that is, from observation today of some data and a model a reconstruction of the past is attempted (this is what is done, for example, in cosmology, when at attempt is made to go back to the first minutes of life of the universe).

The interactions between mathematics and biology represent contemporaneously a challenge and an opportunity, both for the mathematicians and for the biologists, in frameworks that are exquisitely multidisciplinary. Some of these contexts, on the basis of the stress placed on specific determinate aspects, are being organised into new disciplines (to cite only a few key words: biomathematics, theoretical biology, computational biology, system biology).

It goes without saying that the contents of this article are far from illustrating all of the possible interactions between mathematics and the life sciences, and biology in particular. The intention is only to recount briefly two recent experiences, just as you would recount small episodes in a story that is longer and continually evolving. In the world of biology the contemporary presence of many spatial and temporal scales are evident: beginning from a description at a molecular (microscopic) and cellular level, to the level of the single individual, arriving finally at the great complexes of ecosystems. We therefore have the presence of various phenomena, from molecular biology to ecology, and even from the point of view of the genome (a discipline that occupies itself with mapping, of sequencing, and of the analysis of genomes in order to study genes and their functions) it is necessary to classify and search for algorithms that are efficient for an enormous quantity of information, while for the study of some populations in ecology it might happen that only small amounts of data can be collected.

In what follows, the leading characters in the various episodes on which I will touch will be cells, that is, those fundamental structural and functional units of living organisms. In particular we will see cells that move, that form structures, cells that communicate, and that process information together. The scale that interests us is an intermediate scale (mesoscale); from the smaller scales at the molecular level we inherit useful indications as to the phenomena in question, but a detailed formal description will not be necessary.

For the bibliographic part it would be senseless to pretend to give exhaustive indications even for specific fields; we will limit ourselves, even though we are aware of the large "non-fulfilment", to recommending the volume by V. Comincioli regarding the applications of mathematics in the applied sciences [1], the book by G. Israel [2] for a more historic view of mathematical models, to the recent, vast work of V. Comincioli on biomathematics [3] and, for an approach aimed at students in the three-year programme of secondary (high) school, the textbooks by A. Quaternoni, F. Saleri, and A. Veneziani [4].

## First Episode. The Meaning of Cells for the Network

Cell migration is involved in many biological phenomena (for example, in embryonic development, where cell migration is essential in the processes of morphogenesis, from gastrulation to the development of the nervous and circulatory systems). Here we want to analyse in particular the formation of the blood vessels, which originate either by a process of vasculogenesis or of angiogenesis. By vasculogenesis is meant the formation of capillary-like vessels beginning with endothelial cells, which are differentiated *in situ* by groups of mesenchymal cells (hemangioblasts or blood islands) in very precocious states of embryonic development (Fig. 1). The endothelia originates from the peripheral component of the blood island, while the blood cells originate from the central component. The endothelial cells divide and fuse to form a primitive vascular plexus that, with the beginning of blood circulation, gives rise to a arterovenous vascular network. The growth of the blood vessels in the developing organs takes place prevalently as a result of a process of invasion of the primitive formations of the organs by the capillaries, beginning with the primitive

85

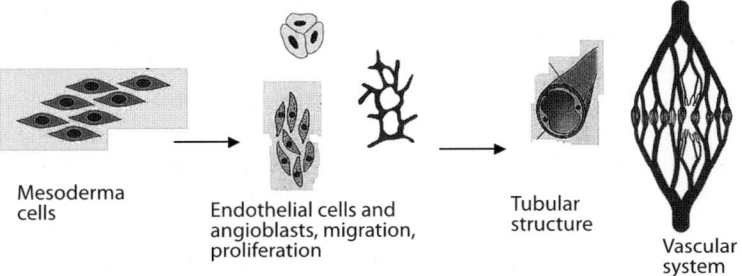

Mesoderma cells

Endothelial cells and angioblasts, migration, proliferation

Tubular structure

Vascular system

**Fig. 1.** Diagrams of some of the developmental phases of vascularization

vascular plexus. The capillary network (responsible for the distribution of nutrients in vertebrates) are characterised by intercapillary distances comprised between 50 and 300 μm, dictated by the necessity of realising an optimal metabolic exchange [5] and the dimension of the structures appears constant even from one species to another.

In contrast, by angiogenesis is meant the generation of new vessels by endothelial cells of pre-existing blood vessels. The study of angiogenesis is of particular importance for the analysis of the growth of tumours. This correlation was initially hypothesised by Folkman (1971). At the beginning of its growth the tumour uses the capillary network of the original tissue both for replenishment of oxygen and nutrients and for elimination of toxic substances. This structure is therefore not sufficient when the solid tumour surpasses a certain dimension (on average $2\,mm^3$); in this case the tumour stimulates angiogenesis by means of suitable soluble factors for the proliferation of new vessels. Here we will deal with only vasculogenesis (and we note right away that there can be different models for angiogenesis).

What mechanism regulates the dimension of the capillary structures? We will limit our attention to a phenomenological description of the process. The process of formation of capillary networks can be partially reproduced in the laboratory [6] and can be observed by suitable videomicroscopic techniques [7]. Through these observations it is possible to confirm that the formation of the vascular network proceeds through three principal stages. In the first phase the endothelial cells (initially spread in the gel of the experimental apparatus) move autonomously and make contact when they meet, forming a continuous multicellular network, whose geometry will not be substantially modified in any essential way during the successive stages. Thanks to the techniques for tracing the trajectories of the individual cells [8], it can further be observed that the movement of the cells is characterised by a high degree of directional persistence, and it is only partly random. In the second stage, the network is deformed and adapted. Finally, the individual cells fold onto themselves so as to form actual capillary tubes along the lines of the network that has formed during the course of the two preceding stages.

Recently, thanks to a multidisciplinary collaboration, a mathematical model to describe the first stage of molecular migration was developed [8], with the consequent formation of the network. Experimentally the motion appears to be essentially along the direction of the highest concentration of cellular matter; this suggested the existence of some form of intercellular communication based on the transmission of chemical signals. This type of directional movement of cells (or organs, or other biological entities) induced by a chemical is called chemotaxis. The mathematical model developed in [7] and [8] is two-dimensional, in that it describes the formation of vessels on the part of the endothelial cells in the laboratory and in the phase of embryonic development for which such a hypothesis is reasonable and realistic. In living organisms the formation of capillary networks occurs either in the first stage of embryonic development or in the adult organism, for example, during the process of the healing of wounds. In this last case, the environment in which the migration takes place is the three-dimensional extracellular matrix, which is "perforated" by the endothelial cells by means of particular proteolytic compounds. During migration the cell is polarised, assuming the elongated form in which it is possible to distinguish clearly a front and back part. This leads again to a model that has to now be "acclimatised" in the three-dimensional space even though it is formally similar to that considered in [8]. Our research group is dealing with three-dimensional cellular migration, both from the point of view of modelling and from that of numeric simulation. The cells are described as a density $n(x, t)$ that moves with velocity $v(x, t)$, which is influenced by the gradients of two chemical factors, one chemoattractive and the other chemorepulsive, represented across scalar fields of concentration $c(x, t)$ (chemoattractive created by the cells themselves) and $r(x, t)$ (short-range chemorepulsive, which turns out to be an inhibitor and whose presence has been confirmed experimentally several times). The temporal variable $t$ varies from 0, the initial time, at a certain time $T$, while the spatial variable $x$ varies from a limited domain of space, thus in $R^3$. We therefore have the following system of partial differential equations:

$$
\begin{cases}
\dfrac{\partial n}{\partial t} + \nabla \cdot (n\mathbf{v}) = 0 \\[2mm]
\dfrac{\partial v}{\partial t} + \mathbf{v} \cdot \nabla \mathbf{v} = \mu(n, c)\nabla c - \nabla p(n) - v\nabla r - \beta \mathbf{v} \\[2mm]
\dfrac{\partial c}{\partial t} = D_c \Delta c + \alpha_c f(n) - \dfrac{c}{\tau_c} \\[2mm]
\dfrac{\partial r}{\partial t} = D_r \Delta r + \alpha_r g(n) - \dfrac{r}{\tau_r}
\end{cases}
$$

In this system the functions $f$ and $g$ and the coefficients $\alpha_r$ and $\alpha_c$ are linked to the production of a chemical agent (repulsive or attractive); the time constants $t_r$ and $t_c$ indicate the time of decay, while the term $\nabla p(n)$ represents a pressure for modelling the short-range mechanical interactions between cells (they get closer but do not penetrate). The function $\mu$ represents the sensitivity of movement with respect to the variation of the concentration of the chemoattractor. The system "rains from the heavens" but we can observe that the first equation is a equation of conservation of total mass over the time of the population of the cells; in the second equation are found all the terms that influence the velocity and its variation (thus with these are described the amoeboid movements of the cells); finally, the last two equations represent the diffusion of the principal chemical factors present. For the system under consideration an explicit solution is not available, and it is necessary to have recourse to a numerical simulation.

In our numerical experiments the spatial domain is represented by a rectangular parallelepiped with periodic border conditions: these are not biological conditions, but they can be assumed for simplicity. An example of simulation is shown in Fig. 2.

For the numeric approximation of the proposed model we had to develop a new code based on some recently introduced techniques, such as relaxed approximation for systems of reaction, diffusion, transportation and the IMEX methods for temporal integration. While for the biological and biomedical applications of the model we

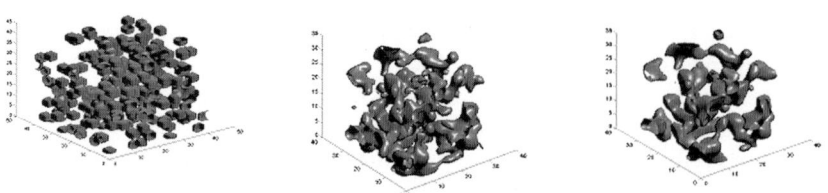

**Fig. 2.** Evolution of cellular density in a numeric simulation of a 3D model of vasculogenesis (*at left*, the initial distribution of cells)

are still at the beginning (as promising as that is), for the parts that are specifically mathematical various problems still remain to be investigated, such as the analysis of the model and of possible states of equilibrium and the stability of the numeric procedure, to cite only two aspects. Will the model turn out to be useful? Certainly. For example, it is important to be able to check the geometry of the vascular network in tissue engineering, which represents a significant and promising line of research and biomedical applications.

Before concluding this first episode, we can make a few observations. First of all, the proposed model is strictly phenomenological and is based on macroscopic in vitro observations related to the migration of endothelial cells. Another approach is based on cellular automata (the famous game of life, still seen on some screensavers, is an example of cellular automata), on the Potts model, and on stochastic rules of evolution [9]. With this kind of approach it is possible to characterise the properties of a single cell and, in particular, it is possible to schematise biochemical mechanisms and determine how variations of these properties can influence the formation of vessels. Figure 3 shows an example of a virtual experiment made with a special cellular automata and a two-dimensional arrangement of cells [10].

The rule of evolution is based on a local evaluation of a minimum of an energy function that depends on the arrangement of the individual cells and their properties (the rule is partly probabilistic).

The second observation regards development of the model to involve other types of chemical agents as well as to proceed to successive stages of the formation of the capillary network. In fact, the proposed model deals with only the first stage of for-

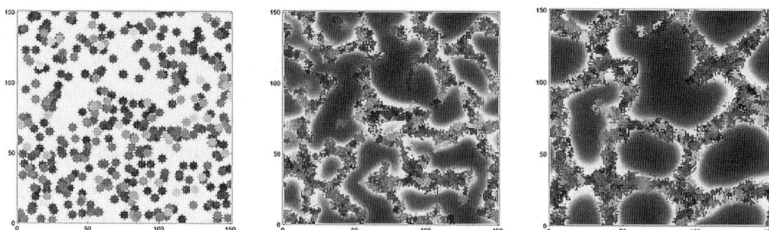

**Fig. 3.** The evolution of a Potts type of cellular automata for vasculogenesis (*at left*, the initial distribution of cells); the rule of evolution is partly probabilistic

mation of the network, while a different approach, and a different model, would deal with a remodelling of the network.

Finally, it is possible to shift the investigation by changing the scale of observation. Recently biological research has shown that in many types of cells the capacity of the cells to orient themselves in the direction of the chemotactic signals is a consequence of the peculiar interactions of some signalling molecules that live in the internal surface of the cellular membrane: the phospholipids PIP2 and PIP3. Starting with the biological data it is possible to construct a kinetic model that describes the initial stages of the polarisation of the cell beginning with a chemotactic gradient [11]. We are at a microscopic level of description, and if we broaden the frame we see that this world disappears and we can observe the population of cells in movement. The analysis and the simulation of the microscopic model might suggest variations of the macroscopic model or might help in the identification of the coefficients (constants or functions) present.

## Second Episode. Cells, Dynamics, and Robots

The exploration of our nervous system, in particular that of the brain, constitutes one of the most advanced frontiers of research; the principal ambition of neuroscience consists precisely in the desire to understand the biological bases of the physiology and the functioning of the brain. To this end disciplines with very diverse traditions conjoin into a single family: from molecular neurobiology to electrophysiology, from biochemistry to genetics, from linguistics to mathematics, from neurophysiology to psychology, from neuroanatomy to biophysics.

Each discipline, even while maintaining its own specificity and methods, constantly interacts with the others. This leads to the increasing awareness that a result obtained in a certain field of investigation can acquire greater meaning only if it is placed in relation to other fields and levels of research. However, in spite of the efforts and continuing progress, still little is known of the brain, and often recourse is made to metaphors that involve the most recent technology. Who hasn't heard the comparison made between the brain and a computer? Certainly there is much more in the brain; yet again the famous warning of Shakespeare's Hamlet comes to mind: "There are more things in heaven and earth, Horatio, than are dreamt of in your philosophy". We won't linger over the immense problems opened up by the relationship between the mind and the brain, fascinating though they are (among the many starting points available we cite the book, even though not recent, of Sir J.C. Eccles, *Facing Reality: Philosophical Adventures of a Brain Scientist* [13]).

In the nervous system the presence of different scales and a certain hierarchy in the systems involved is evident: it goes from molecular phenomena to the description of the biophysics of the cellular membrane to the behaviour of the individual neuron (we have some $10^{12}$ in our brain) and of the connections between neurons (some $10^{16}$ in the brain), from the dynamics of the network of neurons to the analysis of the parts of the brain, to arrive at perception and behaviour.

89

We have mentioned neurons, networks, brain; let us now backtrack and take up our story again. The neurons, or nerve cells, together with the glial cells, are the fundamental "building blocks" of the nervous system in general and of the brain in particular (see [13] and [14]). With respect to other cells, with which they share general organisation and the biochemical apparatus, the neurons possess unique characteristics, including the characteristic shape of the cell, an external membrane that is capable of producing nervous impulses and the synapses (a structure that allows the transfer of information from one neuron to another). Even though there are neurons with different morphologies, there exist common structural characteristics: the cellular body or soma, the dendrite and the axon (see Fig. 4).

The cellular body contains the nucleus and the biochemical equipment necessary for the synthesis of the enzymes and other molecules essential for the life of the cell (normally the body of the cell is approximately spherical or pyramidal).

The dendrites are fine, tubular-shaped extensions that tend to subdivide to form a branching structure around the cell body; the dendrites constitute the principal physical structure for the reception of "arriving signals"

The axon can extend for a remarkable distance from the cell body and furnishes the structure along which the signals travel from the cell body of a neuron towards other parts of the brain and nervous system. The signals that are exchanged and in a certain sense processed by the neurons are electrochemical signals. The work that the neurons do is made possible by the properties of the membrane that covers their entire structure. A stretch of membrane is composed of a double layer of phospholipids (insulators), in which protein channels (conductors) open to allow specific ions to pass. The channel is composed of various protein subunits that permit the channel itself to pass from a closed to an open state and vice versa (this process may require the passage through intermediate states).

The channel has to be able to respond to specific stimuli; further, the greater part of the channels possess a sort of sensor of potentiality that permits them to open up following a depolarisation, that is, a modification of the transmembrane potential

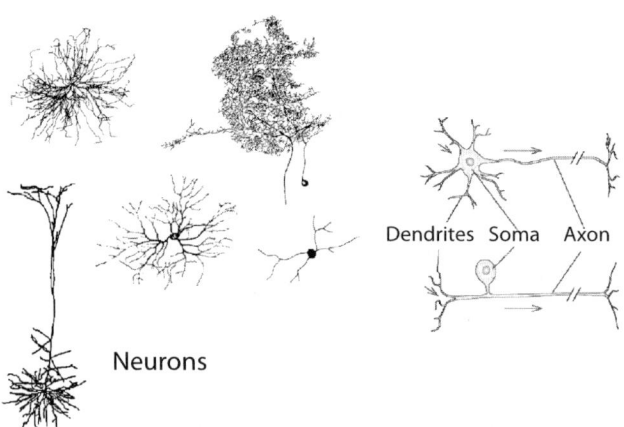

Dendrites  Soma  Axon

Neurons

**Fig. 4.** Neurons with various morphologies (*at left*) and basic organisation of a nerve cell (*at right*; the *arrows* represent the direction travelled by the signal through the neuron)

from the resting value, which is on average comprised between $-70\,\text{mv}$ and $-80\,\text{mv}$, to higher values. The propagation of a nervous impulse along the axon coincides with a localised flow through the membrane of sodium $Na^+$ ions entering, followed by an exit flow of potassium $K^+$ ions. The impulse starts with a slight depolarisation through the membrane; this slight variation induces an opening of some channels of sodium, with a consequent increase of the potential; the flow of sodium ions entering increases until the internal surface of the membrane is locally positive. The inversion of potential, which starts from a negative resting value, closes the sodium channels and opens those of potassium, re-establishing the negative potential: thus is created the propagation of an action potential (Fig. 5).

The action potential is therefore an electric impulse that propagates along the neural membrane.

The potential of action is propagated through the axon, thus the signal is passed to other neurons through the synapses and the dendrite tree. According to an approach introduced by A.L. Hodgkin and A.F. Huxley (winner of the 1963 Nobel Prize for Medicine) and of other researchers in the 1950s, an equivalent electric circuit can be used to describe the phenomenon, one in which the lipid component can be represented by a condenser and the ion channels by non-linear conductances connected to a battery. We indicate the current due to a certain ion channel with $I_k$; thus:

$$I_k(t) = (V - E_k)\, g_k(t, V)\,,$$

where $V$ is the transmembrane potential, $E_k$ the Nernst equilibrium potential, $g_k(t, V)$ the conductance. For a unit of length, or of area, the conductance $g_k(t, V)$, due to the presence of a certain population of channels, can be expressed as the product of a maximum conductance $g_{max}$ for the fraction of open channels. This fraction of open channels is determined by hypothetical variables of activation and disactivation usually indicated by $m$ and $h$ respectively. In general, then, we have:

$$g(t, v) = g_{max} m(t, V)^p h(t, v)^q\,,$$

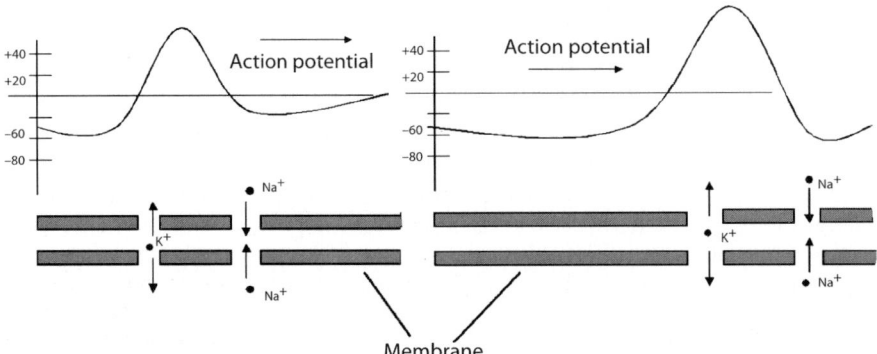

**Fig. 5.** The dynamics of the potential of action (electric signal that propagates along the neural membrane of the axon)

where $p$ and $q$ are whole non-negative constants that model the components of the channel. The dynamics of the variables $m$ and $h$ obey a first-order kinetic, which leads to a system of differential equations; the parameters of the system can be identified experimentally. In particular, the properties of the membrane are deduced in the laboratory by means of the technique of the *patch clamp*, in which blocking the difference in potential of a small area of the cellular membrane serves to isolate the ion channels and furnishes data about the cellular signals [15]. For each membrane segment one arrives at an equation of this type:

$$r\partial_t V - \lambda^2 \partial_{xx} V - I_{\text{ion}} + I_{\text{app}} = 0$$

and for all couples $(m,h)$ we have differential equations that describe their dynamics. Given that for each segment of the individual neuron we have relations similar to those written above, what is obtained is a system of equations of non-linear diffusions coupled with the non-linear dynamics of the ion channels, and with appropriate boundary conditions (of connection) between the various segments of the neuron. This is what happens for the individual neuron; but how are various neurons interconnected among themselves? What "language" is used to communicate?

What relationship is there between schemes of interconnection, neuron dynamics and behaviour?

The connections between neurons are established thanks to the synapses, which represent the points of connection in which the cellular signals are transferred: from the axon of one neuron to the dendrite of another. There exist chemical synapses (the most representative of the nervous system) and electric synapses, which are rarer and permit a rapid propagation of depolarisation. In chemical synapses the neuron that transfers the nervous impulse (presynaptic neuron) terminates with the synaptic button, which contains numerous vesicles full of specific molecules of chemical transmitters (neurotransmitters), which at the arrival of the nervous impulse are released into the space comprised between the two neurons (synaptic fissure) and connects with specific receptors on the postsynaptic membrane, transmitting the nervous impulse to the postsynaptic neuron. The neurons communicate through sequences of brief stereotyped variations of the potential of the cellular membrane (spike, or action potential). Recent experiments have shown that the spatio-temporal structure of this train of impulses plays a fundamental role in the codification of sensory stimuli in the processing of information in the brain and in the processes of learning. The neurons of our brain connect and form networks of cells that constitute the various zones of the brain itself (see Fig. 6). Here we will treat the cerebellum, or "little brain", that is, the area that is found in the back part of our cranium, above the brain stem and under the two great cerebral hemispheres.

For a long time the cerebellum was considered to be only the coordinator of body movements; today, on the other hand, there is evidence that it is also active during the performance of various cognitive and perceptive activities. As does the cortex of the human brain, so too the cerebellum folds onto itself numerous times, compacting an enormous number of nerve circuits in a restricted space. The basic characteristics of these nerve circuits have been known since the first works of

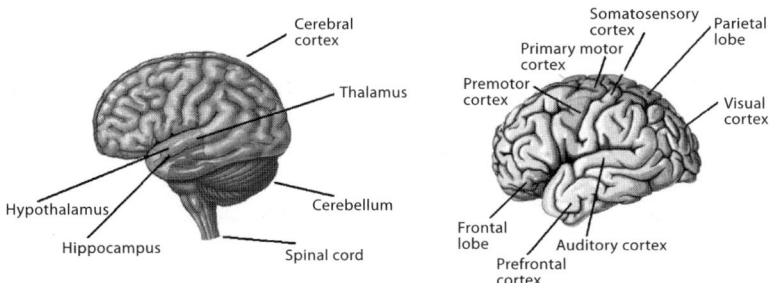

**Fig. 6.** Simple diagram of the organisation of the brain and of the cerebral cortex

the neuroanatomist Santiago Ramón y Cajal, which go back to the end of the 1800s, and of the physiologist Camillo Golgi. The cerebellum has a fundamental role in the functions of coordination and motor control, posture, and balance. In particular, its extended cortical part (the cerebellum cortex) is important for the integration of sensory information and motor commands, and for the capacity to make sensory and motor predictions on brief temporal scales. The cerebellum is able to integrate information coming from different sensory channels and as a consequence permits the processing of a useful representation for adaptive sensory-motor learning. From an anatomical point of view, the construction of such a representation is permitted by the enormous number of granular cells in the cerebellum, which receive the greater part of the afferent signals of the cerebellum cortex (by means of muscoid fibres). The efferent signals of the granular cells subsequently converge in the Purkinje cells, which are present in a much smaller quantity, which produce the signal exiting from the cerebellum cortex. The connection of granular cell to Purkinje cell is presumed to be the principal site where the association between sensory context and motor action is memorised. The Purkinje cells receive an additional afferent signal from the climbing fibres, which are believed to play an important role in the formation of such memories (signals of learning) (Fig. 7).

The cerebellum uses various strategies for sensory and motor codification and learning. These strategies can be transferred, at least in part, to artificial systems of neural networks in order to permit verification of biological hypotheses. Of particular relevance for the processing of information of the cerebellum is the fact that the granule layer, which collects the greater part of the afferent signals that reach the cerebellum cortex, is constituted of a enormous number of small cells, called granulars ($10^{11}$ in humans); the multimodal information coming in constitutes a complete representation of the sensory and motor context according to the classic theory of Marr (1969) and Albus (1971), recodified in a loose representation (that is, one characterised by a percentage of active granular cells much lower than the percentage of muscoid fibres active at the same time). This codification permits the improvement of the capacity to discriminate between very similar sensory contexts.

As a first step, it is interesting to characterise the dynamics of granule cells and their synapses: from the experimental point of view through patch-clamp experiments (see, for example, [15–17], from the modelling point of view the theory of

93

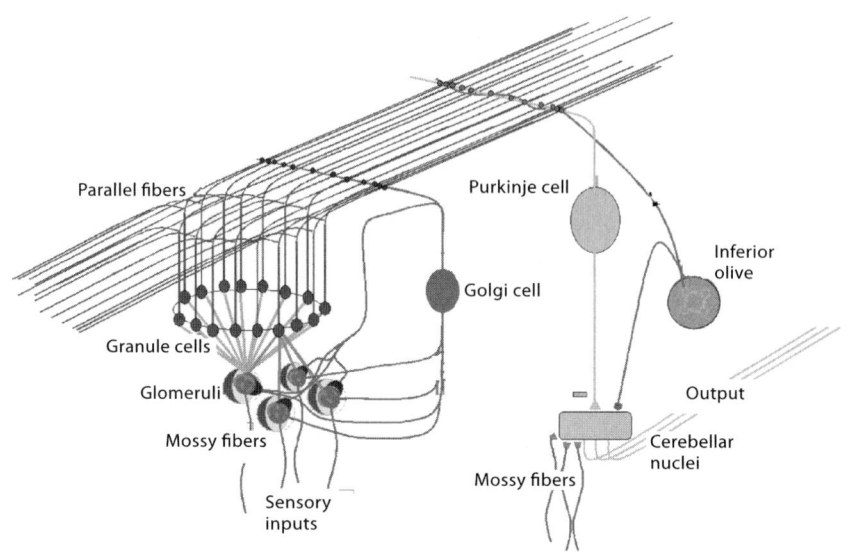

**Fig. 7.** Schematisation of the principal neural circuits of the cerebellum

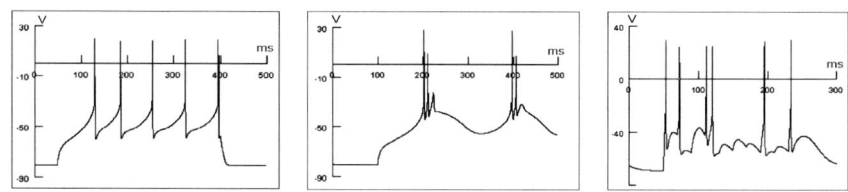

**Fig. 8.** Behaviour of the potential of action under various stimulations: *firing* (*at left*) with periodic repetitions; *bursting* (*center*) with a phase of activity followed by phases of rest; stochastic dynamics (*at right*) obtained by a stimulation with a random signal

Hodgkin and Huxley for the description of the channels, the theory of Rall for the description of the passive properties, and the compartment models, with which the entire neuron is subdivided into zones with homogenous traits (in this way a spatial discretization is created). Figure 8 shows the behaviour of the potential of action of the granule cells under various conditions.

These numeric simulations should be compared with experimental data, and permit an increase in the biological comprehension of the function of the granule cells. With regards to the theoretical approach of the processes of the granule cells, it should be observed that recently techniques are being adopted that are typical of information theory. Measurements tied to entropy and to mutual information are being studied in this sense [18]. Ulterior research activity is focused on synapses and their descriptions.

With this we have concentrated "only" on a single, small cell: the granule cell. All of the deductions drawn should be integrated within the interior of the cerebellum network. Awareness of the anatomic architecture of the neural network of the cerebellum and of the dynamics of the interactions between its components has permitted the development of a complete model of the cerebellum. For the numeric simulation of such a model, however, it is necessary to proceed to a simplification of the description of the cells involved (otherwise an enormous system of differential equations that are difficult to manage is rapidly created).

This model of the cerebellum has been used to create a system of motor control, with the aim of studying the planning and control of complex, precise movements for application to autonomous robots within the European project *Spikeforce*.[1]

The model is related to the control of an artificial arm and comprises a model of the cerebellum capable of making predictions about the sensory and motor contexts; a system of control based on a heuristic learning algorithm; a simulated arm and a simulated visual system that guarantee feedback to the cerebellum.

This has given just an indication of possible applications, but certainly one dream is opening up: the possibility of being able to contribute to the study of possible diseases of the nervous systems.

## Epilogue

We have run quickly through two stories, but certainly a deeper look could reveal immense panoramas.

> The universe (which others call the Library) is composed of an indefinite and perhaps infinite number of hexagonal galleries ... When it was proclaimed that the Library contained all books, the first impression was one of extravagant happiness. All men felt themselves to be the masters of an intact and secret treasure. There was no personal or world problem whose eloquent solution did not exist in some hexagon. The universe was justified, the universe suddenly usurped the unlimited dimensions of hope ... As was natural, this inordinate hope was followed by an excessive depression. The certitude that some shelf in some hexagon held precious books and that these precious books were inaccessible, seemed almost intolerable ... We also know of another superstition of that time: that of the Man of the Book. On some shelf in some hexagon (men reasoned) there must exist a book which is the formula and perfect compendium of all the rest: some librarian has gone through it and he is analogous to a god ... How could one locate the venerated and secret hexagon which housed Him? Someone proposed a regressive method: To locate book A, consult first book B which indicates A's position; to locate book B, consult first a book C, and so on to infinity ... In adventures such as these, I have squandered and wasted my years.
>
> J.L. Borges, 1941, The Library of Babel

95

---

[1] *For more information see the website dedicated to the project, http://www.spikeforce.org.*

As was said at the beginning, the possibilities for cooperation between mathematics and biology are both a challenge and an opportunity. It is certain that in this context the mathematician will recall (or will at least hope) what Leonardo, whose greatness is beyond doubt, wrote:

> … O students, study mathematics, and do not build without foundations … Whoever condemns the supreme certainty of mathematics feeds on confusion, and can never silence the contradictions of the sophistical sciences, which lead to an eternal quackery.

> Leonardo da Vinci, *Windsor Codex*, Royal Library

## Bibliography

[1] V. Comincioli (2004) Metodi Numerici e Statistici per le Scienze applicate, e-book, Università degli Studi di Pavia, http://www.multimediacampus.it.

[2] G. Israel (1996) La visione matematica della realtà. Introduzione ai temi e alla storia della modellistica matematica (new edition), Laterza, Rome-Bari.

[3] V. Comincioli (2005) *BIOMATEMATICA. Interazioni fra le scienze della vita e la matematica*, new edition in e-book format, Apogeo, Feltrinelli, Milano, http://www.apogeonline.com.

[4] A. Quarteroni, F. Saleri, A. Veneziani, *La modellistica matematica va a scuola*, (in preparation).

[5] P.D. Chilibeck, D.H. Paterson, D.A. Cunningham, A.W. Taylor, E.G. Noble (1997) Muscle capillarization, O2 diffusion distance, and V,&O2 kinetics in old and young individuals, *Journal of Applied Physiology* 82, pp. 63–69.

[6] D.S. Grant, K. Tashiro, B. Segui-Real, Y. Yamada, G.R. Martin, H.K. Kleinman (1989) Two different laminin domains mediate the differentiation of human endothelial cells into capillary-like structures in vitro, *Cell* 58, pp. 933–943.

[7] A. Gamba, D. Ambrosi, A. Coniglio, A. de Candia, S. Di Talia, E. Giraudo, G. Serini, L. Preziosi, F. Bussolino (2003) Percolation, morphogenesis, and Burgers dynamics in blood vessels formation, *Physical Review Letters* 90, 118101.

[8] G. Serini, D. Ambrosi, A. Gamba, E. Giraudo, L. Preziosi, F. Bussolino (2003) Modeling the early stages of vascular network assembly, *EMBO J.* 22, pp. 1771–1779.

[9] R.M.H. Merks, S.A. Newman, J.A. Glazier (2004) Cell-Oriented Modeling of In Vitro Capillary Development, in P.M.A. Sloot, B. Chopard, and A.G. Hoekstra (Eds.), *ACRI 2004*, LNCS 3305, pp. 425–434, Springer-Verlag, Berlin Heidelberg.

[10] A. Bravi (2006) Undergraduate thesis, Department of Mathematics, University of Milan.

[11] A. Gamba (2004) Phase ordering model of directional sensing in the eukaryotes. *It. J. Biochem* 53, p. 68.

[12] C. Koch, I. Segev (1998) *Methods in Neural Modeling*, MIT Press.

[13] Sir J.C. Eccles (1970) *Facing Reality: Philosophical Adventures of a Brain Scientist*, Springer-Verlag.

[14] M.A. Arbib, P. Érdi, J. Szentagothai (1998) *Neural Organization*, MIT Press.

[15] E. Neher, B. Sakmann (1992) The patch clamp technique. *Scientific American* 266, p. 44.

[16] E. D'Angelo, T. Nieus, A. Maffei, S. Armano, P. Rossi, V. Taglietti, A. Fontana, G. Naldi (2001) Theta-frequency bursting and resonance in cerebellar granule cells: experimental evidence and modeling of a slow K+-dependent mechanism, *Journal of Neuroscience* 21, pp. 759–770.

[17]  E. D'Angelo, P. Rossi, D. Gall, F. Prestori, T. Nieus, A. Maffei, E. Sola (2004) Long-term potentiation of synaptic transmission at the mossy fiber – granule cell relay of cerebellum, *Progress in Brain Research* 148, pp. 71–80.

[18]  M. Bezzi, A. Arleo, T. Nieus, O. Coenen, E. D'Angelo (2006) Quantitaive characterization of information transmission in a single neuron, In: F. Alexandre, Y. Boniface, B. Girau, and N. Rougier (eds.), *Proceedings of NeuroComp Conference*, pp. 134–136.

# Surprising Coincidences and Some Misunderstandings About "Rare" Events*

Fabio Spizzichino

How many different meetings will streets and squares of a city witness every day? Friends, family, colleagues, rivals, …meet in appointments of many different kinds. These different appointments are, however, often fixed in the same places, places that are attractive, or simply convenient and natural, well-established by some tradition. In such places, one is not surprised to witness the occurrence of several meetings, meetings fixed at the same time, but independently. But which of these are the *coincidences* that cause a state of "surprise", and how justified is the astonishment that one feels when they occur? It is this topic that we consider here in some detail, and illustrate using examples based on elementary probabilistic models.

The subject of coincidences, though fascinating for its philosophical, psychological and literary implications, can give rise to misunderstandings. In many cases these misunderstandings concern questions of probability, and, in particular, questions of events with very small probability. The simple cases we illustrate here can serve as a starting point for interesting reflections.

Much has been written on the subject of coincidences, in different fields and from different perspectives. The subject is complex, and can be analysed from various points of view. Risking complete answers would be daring, but even summing up in a few pages the essential aspects of the literature and of the various contributions to the point seems bold. We will limit ourselves here to some schematic observations of a general character.

First of all it should be made precise what exactly can be understood as a *surprising coincidence*. It is not worth trying to give a rigorous definition; it shall suffice to immediately give an example and to sketch further clarifications in what follows. Many examples and various comments can be obtained simply by an internet search of the word *coincidence*. One of the first hits which result will be

---

* *Translated by Sarah Wolf*

known to many readers, it concerns some "analogies" between Abraham Lincoln and John F. Kennedy:

> Lincoln was first elected to the Congress of the United States in 1846 and Kennedy in 1946. Lincoln was elected President of the United States in 1860 and Kennedy in 1960. Both their successors were called Johnson (Andrew Johnson, born in 1808 and Lyndon Johnson, born in 1908). Lincoln's secretary was called Kennedy and the name of Kennedy's secretary was Lincoln. John W. Booth, Lincoln's assassin, was born in 1839, while Lee H. Oswald, presumed assassin of Kennedy, was born in 1939. Both presidents were assassinated in the presence of their wives and on a Friday: Kennedy was shot in a Ford car, a Lincoln limousine, Lincoln was shot at Ford's Theater, etc.

Coincidences of some kind or other have occurred to everybody (premonitory dreams, fortuitous encounters, numbers with a meaning, and so on). A class of coincidences that we might also have experienced personally go under the name of *serendipity*: cases in which one accidentally finds something that reveals to be of importance later, but that one was not even looking for to being with. There have been remarkable cases of this kind for example in the history of scientific progress [1].

Returning to the meetings in a city, I will report an episode that took place some years ago:

> One cannot exactly say that the corner of two streets, Via A. A. and Via della B., in the quarter "Balduina" in Rome is a place that is common for fixing appointments, but it is here that every Tuesday at three o'clock in the afternoon Giancarlo S. passes by car to pick up his friend Tommaso to go meet a group of other friends for a weekly sports activity.
>
> S. is a rather rare last name in Italy and very rare in Rome: in the urban telephone book, apart from Giancarlo S. you would find just a couple of other families with the same last name. These are however families that he himself does not even know.
>
> One Tuesday Tommaso is unusually late for the appointment and Giancarlo stops to wait for him patiently. After he has waited for a rather long while, another car approaches, two unknown people get off and address him with a friendly "Good afternoon. Are you Mr. S.?"
>
> "Yes, of course, that is me. What has happened?" asks Giancarlo, almost certain at this point that Tommaso is sending them to inform him of some unforeseen event. "What has happened?' – answers one of the other two – "Nothing, we are here for the appointment that we made with you yesterday." With a few words of explanation Giancarlo comes to know that a stranger with the same last name has made an appointment with these two people at the same place, on the same day and more or less at the same time of his own appointment. And moreover, Giancarlo would never have known about it if

Tommaso, who usually is always on time, had not been late precisely this one time.

Some hours later, Giancarlo, after having told me this episode on the phone, asks almost literally: "You, as an expert in probability, what can you tell me about this coincidence?"

What could I have told him? Together we commented on the episode, and I made a few generic remarks; later on I happened to think about the story again several times, and also about other analogous episodes, and the answers to these questions I could have given …

However, instead of concentrating on possible answers, let us consider what is actually highlighted by the questions themselves:

- the episode has been perceived as a *surprising coincidence*,
- the peculiarity (or, let us say the effect of astonishment) of a coincidence is associated, in the current culture, with the observation of events of an *a priori* very small probability.

It is well-known that studies on topics connected to coincidences have been developed by Kammerer, Jung and Pauli; some illustrations of these studies can be found for example in the references [2] and (in Italian) [3]. A number of contributions stress the probabilistic and statistical aspects of the subject more specifically. Some are interesting from the mathematical point of view and thorough in the technical aspects (see in particular [4]; a partial and popular illustration of this subject is also contained in [5]).

The topic of fortuitous coincidences also brings us back to metaphysical discussions about predetermination and free will, existence of *destiny* and so on, and is often associated with the esoteric or paranormal experiences typical of a culture that we perceive as oriental. Besides, this subject has been popularised in the last years by the diffusion of the "New Age philosophy".

Taking a closer look one realizes that stories and variations on this topic are recurrent in the literary production of all of Western culture; consider drama (perhaps even more so in classical than in modern theater), the prose of all times (in particular contemporary prose), and cinema. In short, surprising coincidences permeate all our literary culture.

Can we, however, say anything substantial when passing from literary creations to scientific analyses or to daily reality?

A relevant aspect in the analysis of a coincidence certainly consists in its inherent psychological character; it is us who, upon recognizing a coincidence, attempt to work out its significance. The phenomenon thus has an essentially subjective character: the same series of facts may not by itself give rise to a coincidence without the presence of a subject that recognizes and labels it as such.

Just to give an example, the occurrence of the set of five winning numbers 52-49-18-25-79 in the Lotto drawing on the Venice wheel[1] will not appear to be significant to most people, however, it may appear as a remarkable event to an imaginary person born in 1952 in Venice, whose family consists of a partner born in 1949, a father born in 1918, a mother born in 1925 and an only child born in 1979.

It is therefore obvious that the simultaneous occurrence of certain events may seem a coincidence to one person but not to another. Also, for the same person, one and the same event may be a coincidence with a given state of information and in a certain psychological state, while it will not be under a different state of information or in a different psychological condition. Suppose, however, that a coincidence has taken place; noticing it, analysing it to determine whether it was random, analysing whether there are reasons behind it, etc., in short, reflecting upon the observed coincidence can be the basis of the development of knowledge. This is true also of scientific knowledge, as was mentioned with regard to *serendipity*. Therefore, a particular problem which emerges here is that of evaluating whether a series of observed coincidences is compatible with the hypothesis of pure randomness or whether it tends to contradict it.

This is the point where considerations of a probabilistic and statistical nature come into play: if one determines that the series of observed coincidences has an extremely small probability under the hypothesis of independence between the various phenomena involved (that is, if the series is evaluated as "significant" from a statistical point of view), the question arises whether an alternative probabilistic model, which provides for some form of dependence, is valid. Here, however, misunderstandings or errors in the evaluation of the magnitude of a probability may occur.

First of all, an obvious fact which nevertheless sometimes escapes our attention should be mentioned: events with "very small" probability occur all the time: in natural phenomena, in games of chance, in the fields of economics, finance and insurance, and, for every one of us, in daily life as well as in larger existential questions.

Among the innumerable examples and possible considerations, let us think of a very simple stylized case: in the Italian national Lotto, 11 independent 5-tuples of numbers between 1 and 90 are drawn (almost simultaneously) without replacement, resulting in one of

$$N = (90 \times 89 \times 88 \times 87 \times 86)^{11}$$

equally probable possible outcomes; thus the (a priori) probability of any outcome is given by $p = 1/N$, and is obviously very small.

---

[1] Translator's note: In the Italian Lotto, twice a week 11 drawings of five numbers from 1 to 90 take place. The device that produces the numbers is known as a "wheel". 10 wheels are associated to cities while the eleventh wheel is the "national wheel". To enter the lottery one places a bet, prior to the drawing. This means specifying one or more wheels (up to all wheels) and combination(s) of numbers (up to a maximum of five) that one thinks will be drawn on one of the chosen wheels.

One has to keep in mind that the computation of probabilities varies with the state of information: in particular, all events have probability 1 by the time one knows for sure that they have indeed happened, while at the same time the probability would have been close to zero for many of them, had they been evaluated at the appropriate previous point in time.

When evaluating the significance of a coincidence it is therefore not sufficient to note that an event with an *a priori* very small probability has occurred. One also has to compare this probability, evaluated under the hypothesis of "randomness", with the probability evaluated under possible alternative hypotheses (given the same state of information), and further to keep in mind the a priori probability under these various hypotheses.

Another type of misunderstanding arises from judging the probability of an observed event to be smaller than it actually is (at the current state of information), due to valuation errors or "psychological" effects, or in estimating the probabilities of other events, those that could have occurred and have not been observed, as being larger than they actually were.

Here, we will essentially limit ourselves to some examples related to this last point. It will however be useful to discuss a few basic notions first. In the following section, when introducing the examples that we analyse, the topic of *occupancy models* will be illustrated briefly. The last section is dedicated to a short consideration about the *foundations of probability*, which are also related to the analysis of coincidences.

## Occupancy Models

In various fields, many applications of probability lead to so called *occupancy models*. These models provide a useful scheme in which to present examples of coincidences. Therefore, this section provides a concise description of this topic (for proofs and an in-depth study see, for example, [6]).

Let us consider *r objects* that are placed in *m* different *sites*; suppose that this is done in a random manner. Let the symbol $N_j (1 \leq j \leq N)$ denote the number of objects that are placed at the site *j*. $N_1, \ldots, N_m$ are therefore random variables such that, in formulæ[2]

$$P\{N_1 + N_2 + \ldots + N_m = r\} = 1 .$$

The way in which the objects are placed into the sites determines in particular the joint probability distribution of the random variables $N_1, \ldots, N_m$, that is, the assignment of the probability

$$P\{N_1 = n_1, N_2 = n_2, \ldots, N_m = n_m\}$$

---

[2] *In words, we will say the probability that the sum $N_1 + N_2 + \ldots + N_m$ takes the value r is 1.*

for each $m$-tuple $(n_1, n_2, \ldots, n_m)$, where every $n_j$ is obviously an integer between 0 and $r$ and all possible $m$-tuples belong to the set

$$A_{m,r} = \{\boldsymbol{n} = (n_1, n_2, \ldots, n_m) | 0 \leq n_j \leq r; \sum_j n_j = r\}.$$

It is known[3] that the number $|A_{m,r}|$ of elements of the set $A_{m,r}$ (so, $|A_{m,r}|$ denotes the *cardinality* of $A_{m,r}$), is given by

$$|A_{m,r}| = \binom{m + r - 1}{m - 1}.$$

The random variables $N_1, \ldots, N_m$ are called *occupancy numbers* and every possible joint distribution of $(N_1, \ldots, N_m)$ is called an *occupancy model*. Therefore, an occupancy model is just a probability distribution on a set of type $A_{m,r}$.

It is not difficult to imagine how occupancy models can play a role in all scientific endeavours, in many applications and in many aspects of everyday experience. In different examples and applications, elements of disparate sorts can be considered objects; and the objects are placed into sites of equally disparate types. Before continuing we consider three examples which allow us to visualize some general aspects of occupancy models, and which will then also provide us with a starting point for a discussion on the topic of coincidences.

## The Birthday Problem

Let us number the days of a (non-leap) year from 1 to 365. Further, consider a group of people and denote by $N_1$ the number of individuals in the group who were born on the first of January, by $N_2$ the number of people born on January 2nd and so on (for simplicity we assume that the births on the 29th of February are not registered).

Consider each person as an object and each day of the year as a site. Then, associating to each person her/his day of birth means viewing $(N_1, \ldots, N_r)$ as a vector of occupancy numbers with values in the set $A_{365,r}$ (where $r$ denotes the total number of registered individuals).

---

[3] *The proof is based on the observation that the generic element $n = (n_1, n_2, \ldots, n_m)$ of the set $A_{m,r}$, can be represented graphically by a drawing of the type*

$$* \, * \, * \ldots * \, * \, * \, | \, * \, * \, * \ldots * \, * \, * \, | \ldots \quad \ldots | \, * \, * \, * \ldots * \, * \, * \, | \, * \, * \, * \ldots * \, * *$$

*with $(m-1)$ bars $|$ and $r$ asterisks $*$: one places $n_1$ asterisks to the left of the first bar, $n_2$ asterisks between the first and the second bar, and so on, up to placing $n_m$ asterisks to the right of the $(m-1)$-th bar. Conversely, to each drawing of this kind corresponds one and only one element of $A_{m,r}$. The conclusion is therefore obtained by observing that there are exactly*

$$\binom{m + r - 1}{m - 1}$$

*possible ways of constructing drawings of this type.*

## "Totocalcio" – Soccer Betting

In the soccer betting game "Totocalcio" the outcome of a *"matchday"*, actually corresponding to one week in the championship, consists of a column of 14 elements belonging to the set of symbols $\{1, X, 2\}$. Every element of the column corresponds to a match in the football championship, where 1 denotes victory of the home team, 2 denotes their defeat and $X$ indicates a draw. 14 previously chosen matches are considered.

Let $N_1$ denote the number of 1's, $N_2$ the number of 2's and let $N_X$ be the number of $X$'s in the winning column of a fixed *matchday*.

Here we can view the matches as objects and the symbols $1, X, 2$ as sites; there are thus 14 objects and 3 sites. The results of a *matchday* will attribute a site to each object and therefore one can view $(N_1, N_2, N_X)$ as the vector of the related occupancy numbers; hence, the probability distribution of $(N_1, N_2, N_X)$ is an occupancy model with $m = 3$, $r = 14$.

## Lotteries

In a lottery $r$ tickets are sold to as many players. Let us suppose that there are $m$ classes of prizes, each one corresponding to a different level of winnings. In this case we can view the tickets (or the players who have bought them) as objects and the prize classes as sites: at the moment of the drawing each ticket is assigned to a class. Usually, $m$ is much smaller than $r$ and the largest class consists of prizes of value 0. However, in the following it will be convenient to consider lotteries in which the prizes are distinct and their number is $r$, that is, there are as many prizes as tickets sold. Notice that in this case the occupancy numbers are not random in that they will all assume the value 1 with certainty; however, the assignment of the particular objects (tickets) to sites (prizes) remains random.

We return now to some general considerations about occupancy models.

Two fundamental occupancy models are known as the *Maxwell–Boltzmann* model and the *Bose–Einstein* model. Both models treat the sites symmetrically, but they differ in the fact that the former describes a situation in which the objects are *distinguishable*, while the latter one emerges in a natural way when the objects are *indistinguishable*. We say objects are *indistinguishable* when the observer cannot perceive any differences between them. To visualize the difference between these cases, think for example of objects such as billiard balls of the same colour, or of those same balls numbered from 1 to $r$.

Next, consider the random experiment that consists in placing, in some random way, the $r$ objects into the $m$ sites. Note that, if the objects are distinguishable, the *elementary* event that will be observed corresponds to the description of the site into which each of the $r$ objects has been placed; this description is expressed by the $r$-tuple $s = (s_1, s_2, \ldots, s_r)$, where $s_k$ indicates which of the $m$ sites has been chosen by the object marked with the number $k$ $(k = 1, 2, \ldots, r)$. This demonstrates in particular that the number of possible elementary events of this kind is given by $m^r$. One can also show that among these elementary events the number of those which give rise to a compound event of the type

$$\{N_1 = n_1, N_2 = n_2, \ldots, N_m = n_m\}$$

is equal to the *multinomial coefficient*

$$\binom{r}{n_r \ldots n_m} = \frac{r!}{n_1! \ldots n_m!}.$$

Let $\Omega_{m,r}$ denote the *sample space*, consisting of all elementary events $s = (s_1, s_2, \ldots, s_r)$.

Let us now consider the case in which each of the $r$ distinguishable objects chooses one of the $m$ sites to be placed in with probability $1/m$ (equal for all sites), independently of the choices made by the other objects. This assumption is equivalent to the condition that each elementary event $s = (s_1, s_2, \ldots, s_r) \in \Omega_{m,r}$ occurs with equal probability, and thus that each elementary event has probability $1/m^r$; we will write

$$P(s) = \frac{1}{m^r}, \quad \forall s \in \Omega_{m,r}.$$

The corresponding joint probability distribution of the occupancy numbers $N_1$, $N_2, \ldots, N_m$ is known as the *Maxwell–Boltzmann* model, and is given by

$$P\{N_1 = n_1, \ldots, N_m = n_m\} = \frac{\binom{r}{n_1 \ldots n_m}}{m^r}.$$

To explain the preceding formula it suffices to observe that, starting out with a *uniform* probability distribution on the space $\Omega_{m,r}$, the probability of the compound event $\{N_1 = n_1, \ldots, N_m = n_m\}$ is obtained simply by dividing the number $\binom{r}{n_1 \ldots n_m}$ of favourable elementary events by the total number $m^r$ of possible elementary events.

On the other hand, in the case where the $r$ objects are indistinguishable, the elementary events that can be observed are just events of the type $\{N_1 = n_1, \ldots, N_m = n_m\}$[4] and thus the space of elementary events coincides with $A_{m,r}$.

In this context, the *Bose-Einstein* model is defined as the model that corresponds to a *uniform* distribution on $A_{m,r}$; one therefore has for each vector $(n_1, n_2, \ldots, n_m) \in A_{m,r}$:

$$P\{N_1 = n_1, N_2 = n_2, \ldots, N_m = n_m\} = \frac{1}{\binom{m+r-1}{m-1}}.$$

Thus, both the Maxwell-Boltzmann and the Bose-Einstein model are described by a uniform probability distribution, that is, by a purely random choice. However, the

---

[4] *That is, we do not have the possibility to distinguish which, but only how many objects are placed into each site.*

first one is obtained from a random choice out of the elements of $\Omega_{m,r}$ while the second is obtained from a random choice among the elements of $A_{m,r}$.

One can prove (an explanation can be found e. g. in [7, vol. 2]) that for distinguishable objects the Bose-Einstein model corresponds to a situation of symmetry between the sites with a positive correlation in the choice of sites by the objects. Let us recall that the Maxwell-Boltzmann model also corresponds to a situation of symmetry between the sites, however, the choices of sites by the objects are independent.

A natural generalization of the Maxwell-Boltzmann model is the *multinomial* model. Again, one considers distinguishable objects and each object still chooses a site independently of the behaviour of the others, but there is no longer necessarily a situation of symmetry between the different sites: the site $j$ ($j = 1, 2, \ldots, m$) is chosen by the generic object with probability $p_j$, where $p_1 + p_2 + \ldots + p_m = 1$. In this case

$$P(s) = p_1^{n_1} \cdot p_2^{n_2} \cdot \ldots \cdot p_m^{n_m}, \ \forall s \in \Omega_{m,r},$$

$$P\{N_1 = n_1, N_2 = n_2, \ldots, N_m = n_m\} = \binom{r}{n_r \ldots n_m} p_1^{n_1} \cdot p_2^{n_2} \cdot \ldots \cdot p_m^{n_m},$$

$$\forall (n_1, n_2, \ldots, n_m) \in A_{m,r}.$$

Another particular class of occupancy models emerges when objects are indistinguishable and a condition of *exclusion* between the objects is imposed: each site can contain at most one object. Introducing this condition and maintaining symmetry between the sites one obtains the Fermi–Dirac model:

$$P\{N_1 = n_1, N_2 = n_2, \ldots, N_m = n_m\} = \frac{1}{\binom{m}{r}},$$

where each of the numbers $n_1, \ldots n_m$ can assume only the values 0 or 1. Obviously, in the case of exclusion one must have $m \geq r$.

A very specific but interesting model is obtained in the case of exclusion with distinguishable objects and $m = r$. One can see that in this case the elementary event in the experiment is a *permutation* $\pi = (\pi(1), \pi(2), \ldots, \pi(r))$ of $\{1, 2, \ldots, r\}$. Therefore, the occupancy model consists in a probability distribution on the set $P_r$ of all such permutations.

In models of the latter type, it is interesting to consider the random variable

$$C = \sum_{k=1}^{r} 1_{\{\pi(k)=k\}}$$

that counts the number of *fixed points*, that is the number of values $k \in \{1, 2, \ldots, r\}$ that are fixed by the permutation.

For an example of this kind of situation, one can think of a lottery in which each player is assigned a prize and the prizes are all different (and each player has only one ticket). To illustrate the meaning of the random variable $C$ let us think of the case where the prizes are numbered from 1 to $r$: in this case $C$ indicates how many players receive a prize marked with the same number that is printed on their ticket.

An even more particular model is the one defined by the *uniform* probability distribution on $P_r$:

$$P(\pi) = \frac{1}{r!} \quad \forall \pi \in P_r$$

In the case of the lottery, this is however a rather natural condition: each prize is associated to any ticket with equal probability.

## Examples of Coincidences: Surprising or Not?

We return to the examples mentioned in the previous section. By considering variations on these examples, and with a few additional remarks, we will illustrate some of the various kinds of misunderstandings and errors of a psychological perspective that can occur in the analysis of observed coincidences.

### The Birthday Problem

Consider again a group of $r$ people, and, progressively numbering the days of the non-leap year from 1 to 365, we denote by $N_1$ the number of people born on the first of January, by $N_2$ the number of people born on the second of January, and so on.

If we assume that the birthday of each person is uniformly distributed over all 365 days and we further suppose that the birthdays are stochastically independent from each other, we obtain that the joint probability distribution of $(N_1, N_2, \ldots, N_{365})$ corresponds to a Maxwell-Boltzmann model with $r$ objects and $m = 365$ sites.

Under this assumption, let us now calculate the probability $p_r$ that all birthdays fall on different days[5]:

$$p_r = P^{(r)}\{N_1 \le 1, N_2 \le 1, \ldots, N_{365} \le 1\} .$$

Obviously, we have $p_1 = 1, p_2 = 364/365, p_{366} = 0$ and this probability is strictly decreasing when viewed as a function of $r$.

An interesting question is which number $r$ is the smallest one with $p_r < 1/2$.

Based on the assumption of independence between the birthdays of different people, for $2 \le r \le 365$ we can write

$$p_r = \frac{364}{365} \cdot \frac{363}{365} \cdot \frac{362}{365} \cdot \ldots \cdot \frac{365 - r + 1}{365} .$$

From this formula one easily obtains $p_{22} > 1/2, p_{23} < 1/2$.[6] Thus, if the group consists of 23 or more people, it is more probable to find some coincidence of birthdays than not finding any.

---

[5] *For an event E in the context of the birthday problem, let $P^{(r)}(E)$ denote the probability that E occurs, having fixed the total number of people equal to r.*

[6] *More precisely, we have $p_{22} = 0.5244$ and $p_{23} = 0.4928$.*

Before working out the computations, this result may seem unexpected; however, (without performing any further computations) it allows us to imagine that, for large enough $r > 23$, it is very likely to observe many birthday coincidences.

If the group consists of people who did not know each other beforehand and who meet, for example, to set out on some adventure together, these coincidences could appear rather surprising and could even appear to be precursors of some hidden meaning, while in fact they are completely compatible with the assumption of independence between different people's birthdays. (If, instead of birthdays one thinks of some other characteristic that is emotionally more closely related to the upcoming adventure, imagine how much bigger the psychological impression drawn from the coincidence might be.)

The coincidence effect (maintaining the assumption of independence) would obviously be amplified if the probability distribution of the birthdays over the different days were not uniform.

From a more technical point of view, many precise approximation results for the birthday problem and related questions can be obtained by means of "Poisson approximations". We shall see a Poisson approximation argument below, when dealing with the number of fixed points in a random permutation. The paper [8] gives a useful exposition about applications of Poisson approximations in many fields, including coincidences.

## "Totocalcio" – The Winning Column

In the soccer betting game "Totocalcio" each gambler predicts the column of outcomes 1, $X$ and 2 before the matches are played; in reality one usually "plays" several (and sometimes many) columns, but for the sake of simplicity, we will assume here that each gambler plays only a single column.

Before going on it is timely to put forward a reminder about the statistical behaviour of the match outcomes: generally the home team is favoured. This is in particular confirmed by the statistical analysis of the outcomes of the matches in the first two Italian leagues "Serie A" and "Serie B" over the last few years, which shows that approximately 50% of the outcomes are 1, about 30% are $X$ and about 20% are 2.

Denoting by $N_1$ the number of outcomes that are 1, by $N_2$ the number of matches resulting in a 2, and by $N_X$ the number of results $X$ in the winning column for a fixed *matchday*, the probability distribution of $(N_1, N_2, N_X)$ is, as already known, an occupancy model with $m = 3$, $r = 14$; note that the "objects" (the matches) can obviously be considered distinguishable. Every gambler will have a personal state of information, expressed by a probability distribution on the set $\Omega_{m,r}$ of possible columns $\mathbf{s}$, with a resulting occupancy model for $(N_1, N_2, N_X)$.

Excluding exceptional situations, it is rather natural to assume independence between the results of the different matches. Thus, to fully describe the state of information of a gambler, it suffices to express his probabilities $p_1(k), p_2(k), p_X(k)$ for the three outcomes, with regard to each match $k$ ($1 \leq k \leq 14$), as

$$P(\mathbf{s}) = p_{s_1}(1) \cdot p_{s_2}(2) \cdot \ldots \cdot p_{s_{14}}(14), \quad \forall \mathbf{s} \in \Omega_{m,r} = \{1, X, 2\}^{14} .$$

Let us now consider four different gamblers **A**, **B**, **C** and **D** with different kinds of information.

**A** is a person who plays only once in a while (for example a tourist), without even getting informed about the rules of "Totocalcio" and about the meaning of the different matches and the symbols $1, 2, X$; let us suppose that for $1 \leq k \leq 14$, gambler **A** assumes

$$p_1(k) = p_X(k) = p_2(k) = \frac{1}{3} .$$

**B** is a gambler who also plays only once in a while, and though he knows the rules of the game and the general statistics of the matches well, he is not well informed on the recent performance of the different teams; **B** might be, for example, an Italian who has been living abroad for many years. Let us suppose that, based only on the statistics, gambler **B** assumes

$$p_1(k) = 0.5, \ p_X(k) = 0.3, \ p_2(k) = 0.2 .$$

**C** is the typical gambler who knows the statistics on all matches well; in addition he always has up to date information on the performance of the different teams and indeed develops his own ideas about the situation, which often constitute the subject of articulate discussions and revisions; in his case $p_1(k)$, $p_X(k)$ and $p_2(k)$ depend strongly on the specific *matchday* and, for a fixed *matchday*, vary distinctly with $k$.

**D** is a rather atypical gambler, who views the matches as indistinguishable objects, with symmetry between the different outcomes. He assumes a Bose–Einstein model for the occupancy numbers $N_1$, $N_2$ and $N_X$ (among other things **D** admits a certain positive correlation between the results of the different matches).

Suppose now that we observe the outcome $(1, 1, \ldots, 1)$: *all home teams win!*

Let us, for short, denote this result by the symbol **E**, and first of all note that it can also be written in the form $(N_1 = 14, N_2 = 0, N_X = 0)$.

How do **A**, **B**, **C** and **D** react to the occurrence of **E**?

**D** does not have anything to say to the point: the probabilistic model he assigned for $(N_1, N_2, N_X)$, as we have said, corresponds to the *Bose–Einstein model*; under this model, the probability of **E** is given by

$$\frac{1}{\binom{m + r - 1}{m - 1}} = \frac{1}{\binom{16}{2}} = \frac{1}{120} .$$

Even if this probability is rather small, **E** is just one of the 120 equally probable elementary events that could be observed.

Neither for **A**, whose state of information corresponds to a *Maxwell-Boltzmann model* for $(N_1, N_2, N_X)$, should there be a reason for astonishment. Even if in his scheme the probability of **E** is extremely small (that is, equal to $1/3^{14}$), **A** can say that one of the $3^{14}$ elementary events has occurred and that any other outcome would have had the same probability (we will return to this point in a little while).

The reaction of **C** obviously depends on his specific forecast; in particular if various supposedly weak teams have played at home against strong competitors, **C** might, with good reason, be very surprised by the occurrence of **E** and would be inclined to imagine some factor common to all matches that would have simultaneously favoured all home teams, contradicting his initial conviction of independence.

The state of information of **B** corresponds to a *multinomial model* for $(N_1, N_2, N_X)$. The possible reaction of **B** when confronted with the outcome **E** is possibly the most illuminating one for our discussion. Let us look at two types of arguments:

i) Since his forecast viewed the home teams as distinctly favoured, **B** should not be surprised by the result. Even though the a priori probability of the resulting outcome is very small $(1/2^{14})$, it turns out, however, to be larger than the probability of any other possible column.

Yet, in **B**'s situation (and even more so for **A**'s) one could also reason as follows:

ii) Written in terms of occupancy models, **E** is equivalent to the event $(N_1 = 14, N_2 = 0, N_X = 0)$; the a priori probability of this event is very small, and a different, more balanced outcome for the occupancy numbers, for example $(N_1 = 7, N_2 = 4, N_X = 3)$, would have had a far larger a priori probability given by

$$\frac{14!}{7! \cdot 4! \cdot 3!} \cdot \frac{1}{2^7} \cdot \frac{1}{3^4} \cdot \frac{1}{5^3}.$$

Hence, an extremely improbable and significant coincidence has taken place: there must have been some correlation which has simultaneously favoured all home teams beyond what was expected under the assumption of independence.

Where is the contradiction between arguments i) and ii)? Instead of dwelling on an analytical explanation, we will give an answer from a practical point of view.

Suppose for the moment that, in order to win at "Totocalcio", a gambler must guess the values of $(N_1, N_2, N_X)$. Then with the state of information of **A**, and also with that of **B** and similarly of **C**, there would not be any doubt that it would be better to choose, for example, the outcome $(N_1 = 7, N_2 = 4, N_X = 3)$ rather than $(N_1 = 14, N_2 = 0, N_X = 0)$. In fact, the event $(N_1 = 7, N_2 = 4, N_X = 3)$ consists of a large number of elementary events, while the outcome $(N_1 = 14, N_2 = 0, N_X = 0)$ consists of just one elementary event (note that for **D** the two results are equally probable). However, in order to win, one is asked to guess in detail the whole column of single outcomes; that is, one has to get right not only *how many* but also *which* matches produce the outcomes 1, 2 and X respectively. Therefore, under **B**'s state of information, and being restricted to bet on only one single column, it becomes preferable to choose the column $(1, 1, \ldots, 1)$ rather than any other.

This situation highlights the following fact: one should not get caught in the snare of confusing the probability of a single column (that is, say, a *microscopic* result) with the probability of a set defined in terms of occupancy numbers (a *macroscopic* result)!

The possibility of confusion between these different situations (which we call *microscopic* and *macroscopic*, respectively) can help explain some of the misunderstandings in the analysis of coincidences, even at a more general level.[7]

## Random Permutations and Lotteries

Let us return to random permutations of $r$ objects, as in the case of a fair lottery with $r$ tickets numbered 1 to $r$, and $r$ prizes numbered 1 to $r$, where each prize is assigned to any ticket with equal probability.

For a given $r$, the number of *fixed points* of our random permutation will be denoted by the random variable $C_r$; what can we say about the probability distribution of $C_r$?

In the case where $r = 2$, only the two equally probable outcomes $(1, 2)$ and $(2, 1)$ can occur, therefore each has probability $1/2$.

The first outcome yields $C_2 = 2$, the second one $C_2 = 0$. Thus, $C_2$ is a random variable that can assume only the two values 0 and 2, each with (equal) probability $1/2$. Its expected value is therefore given by:

$$E(C_2) = \frac{1}{2} \times 0 + \frac{1}{2} \times 2 = 1 .$$

In the case $r = 3$, the six possible outcomes are $(1, 2, 3)$, $(2, 1, 3)$, $(1, 3, 2)$, $(3, 2, 1)$, $(2, 3, 1)$ and $(3, 1, 2)$, and therefore each occurs with probability $1/6$.

---

[7] *Let us examine the observation of R different phenomena and for simplicity let us introduce the following framework: the first phenomenon can arise in $h_1$ different ways or modes $m_1(1), \ldots, m_1(h_1)$, the second phenomenon can present itself in the modes $m_2(1), \ldots, m_2(h_2), \ldots$, the R-th phenomenon in the modes $m_r(1), \ldots, m_R(h_R)$.*

*Suppose also that in our personal psychology there exist correspondences between some modes of different phenomena; let these match in the following way: $\{m_1(1), m_2(1), \ldots, m_R(1)\}, \{m_1(2), m_2(2), \ldots, m_R(2)\}, \ldots$*

*That is, we consider the case where we would perceive the simultaneous occurrence of $\{m_1(1), m_2(1), \ldots, m_R(1)\}$ or of $\{m_1(2), m_2(2), \ldots, m_R(2)\}$ and so on as peculiar coincidences.*

*Let there be $m - 1$ classes of this type; some modes will not belong to any class.*

*This situation can subconsciously determine occupancy numbers $N_1, N_2, \ldots, N_{m-1}, N_m$, where $N_1$ is the number of phenomena that present themselves in class 1, ..., $N_{m-1}$ is the number of phenomena that occur in class $m - 1$, $N_m$ is the number of phenomena that take place in modes which do not belong to any class.*

*When an outcome has "balanced" occupancy numbers, one may not realise that one is effectively observing a microscopic result, and may only become aware of it when, for some $j$ the number $N_j$ takes a much larger value than the others.*

*A further "coincidence effect" can come about because one is not aware of observing some phenomena at all, until they occur in the mode that belongs to a class with a large value of the associated occupancy number; that is, under the R phenomena one only notices those which take place in a mode that has occurred with high frequency.*

Now, for $(1, 2, 3)$ we have $C_3 = 3$; the outcomes $(2, 1, 3)$, $(1, 3, 2)$, $(3, 2, 1)$ yield $C_3 = 1$; and finally, in the cases $(2, 3, 1)$ and $(3, 1, 2)$ we have $C_3 = 0$. Thus $C_3$ is a random variable that can assume only the values $0$, $1$ and $3$, with respective probabilities $1/3$, $1/2$ and $1/6$ and hence its expected value is given by:

$$E(C_3) = \frac{1}{3} \times 0 + \frac{1}{2} \times 1 + \frac{1}{6} \times 3 = 1 .$$

One can easily show that, for any $r$, the expected value of $C_r$ remains equal to $1$.[8]

On the other hand, the probability distribution of $C_r$ obviously varies with $r$, but one can prove that it converges rapidly to a Poisson distribution with expected value $1$, that is, with increasing $r$ we have

$$P\{C_r = n\} \rightarrow \frac{e^{-1}}{n!} ,$$

and in particular

$$P\{C_r \geq 1\} \rightarrow 1 - e^{-1} .$$

We return to the subject of lotteries. Often it may happen that one plays in a lottery of a rather special kind, perhaps even without buying a ticket. Simplifying, we can say that there is a group of $r$ players and that each of them will receive a prize, but the value of the prize is "subjective": for each player there is just one prize that he considers "valuable" (or interesting, remarkable etc.) and all other prizes have no value for him. Suppose here for simplicity that the prize considered valuable is different for each player. Let both the prizes and the players again be numbered 1 to $r$, and more precisely, let the prize with the number $k$ be the one that is judged fortunate by player $k$. Any random assignment of the prizes determines a random permutation, and the number of "winners" (players who feel lucky) coincides with the number of fixed points $C_r$.

Now compare the lottery described above to the standard one, in which only a single prize is valuable and all others have no value; let us denote these lotteries as lottery $b$ and lottery $a$, respectively. From the point of view of the player there is no difference between the two cases: his probability of "winning" is equal to $1/r$ in both cases.

From the "collective" point of view however, there is a difference: in case $a$ the number of "winners" is equal to one with certainty; in case $b$ the number of winners is a random variable, which, for the previously mentioned reasons has expected value $1$, and its probability distribution is approximately Poisson.

A particular psychological effect that can occur a posteriori for the winner of an enormous prize in a lottery of type $a$ has often been noticed: he will recall everything that has happened to him before buying the ticket and he will notice a series

---

[8] *Basically, this follows from the fact that there are r elements and each of them has probability $1/r$ of giving rise to a fixed point. The desired conclusion is obtained using the linearity property of the expected value.*

of circumstances that have brought about buying precisely this ticket: he may feel a kind of "destiny effect".

We can imagine that, in the lottery of type $b$, this effect may be noticed in an even more acute manner, due to the randomness in the existence of a winner. This effect, however, is not justified: in fact, at the individual level, there is no difference at all between the two situations.

At the collective level one could have an effect of "surprising coincidence" in observing various "fixed points". And one may be led astray by the intuitive impression that the probability of a given number of fixed points would depend on the total number $r$. This impression would however not be justified, since the probability distribution of $C_r$ varies little with increasing $r$, as was noted earlier.

## Analysis of Coincidences and the Foundations of Probability

Various different meanings can be associated to the term *probability* and any ambiguity, naturally, can give rise to misunderstandings. A very short discussion about some of these meanings and their mutual relations will be useful in view of the connections to the subject of coincidences that arise here.

The mathematical meaning of the term *probability* is generally taken to be the one formalized in the axiomatic theory founded by Kolmogorov [9], where the probability $P$ is a (countably additive) measure on a $\sigma$-algebra of subsets of a space $\Omega$ (*sample space*) with $P(\Omega) = 1$ (axiom of *normalization*). For a textbook on the topic see, for example, [10]. See also the chapter by Carlo Boldrighini in this book for an interesting historical note about the axiomatization of probability.

The elements of $\Omega$ are "interpreted" as the *elementary events* that can be observed in a random experiment. In the case where $\Omega$ is a finite set (say, $\Omega = \{\omega_1, \omega_2, \ldots, \omega_N\}$), any subset of $E$ of $\Omega$ can be seen as an event (known as a *compound* event if it contains more than one element) and, as an immediate consequence of the axiom of additivity, the probability $P(E)$ equals the sum of the probabilities of the elementary events (elements of the space $\Omega$) of which $E$ is composed.

A very particular case, one however of fundamental importance, is the case where in addition to $\Omega$ being a finite set, a situation of *symmetry* between the different elementary events emerges naturally, a situation where all elementary events are equally probable.

In that case, an immediate consequence *of* the axiom of normalization is that:

$$P(\omega_i) = \frac{1}{N}, \; i = 1, 2, , \ldots, N$$

and for any subset $E \subseteq \Omega$, due to the axiom of additivity,

$$P(E) = \frac{|E|}{N},$$

where $|E|$ denotes the cardinality of $E$.

Such a situation occurs naturally in many applications (in particular in the field of physics, genetics, gambling etc.) and the preceding formula is translated into natural language by the well known statement:

– *the probability of an event E is given by the ratio of favourable cases to possible cases.*

In this context the computation of probabilities basically reduces to questions of *combinatorial analysis*; the mathematical theory derived (essentially elementary, but nevertheless often quite complex in practice) finds an immediate correspondence in applications, which in turn correspond perfectly to the theory. In this case we are in the context of *combinatorial* probability or of *classical* probability, which forms the ideas of probability on which the first developments of the theory, from 1500–1600 onwards, were based.

Even for a finite sample space $\Omega$, the situation changes considerably in cases where imposing the condition that all elementary events have equal probability would be unrealistic.

When considering applied problems, especially of a statistical kind, in some cases one can be forced to express what is effectively meant by *probability* in a specific question; that is, one has to specify how the probabilities of various events are interpreted and "measured" (yet maintaining the mathematical behavior and the "rules" of probability theory as codified by the axioms of Kolmogorov). This situation differs from the one of combinatorial probability, in which one might say "I do not know what probability means "physically", but I **do know** that *it is equal* for all elementary events". At this point many different interpretations with different rationales regarding measurement can arise. These differences are often determined by the kind of problem and the concrete application at hand.

Problems of this sort are perhaps even more relevant in cases where $\Omega$ is not a finite space. One thus arrives at the "foundations of probability". As is well known, there exists a vast and very profound debate on these problems, one rather difficult to summarize, or be aware of in its fullest extent.

It is fair to say, however, that in scientific thinking and applications, as well as in many other endeavours, two fundamental lines in the interpretation of probability have emerged: one ("frequentist") that measures probability based on certain observed frequencies of success, and one ("subjectivist") which connects probability to the mental organization and to the behaviour of an individual faced with a situation of uncertainty, or in other words, a situation with an incomplete state of information.

Apart from the axiomatic definition we thus have three different ways of "interpreting" the probability of an event: a combinatorial interpretation (ratio of favourable to possible cases), a frequentist interpretation (ratio between the number of successes and the number of trials, with a large number of "independent" analogous trials); and a subjectivist interpretation (degree of belief in the occurrence of an event, as assessed by an individual).

These three interpretations essentially differ in their respective fields of applicability:

- the combinatorial interpretation is valid only in the case of a finite sample space with symmetry between elementary events
- the frequentist interpretation puts aside such restrictive conditions, but nevertheless requires that the events to which one wants to assign a probability be "repeatable"
- the subjectivist interpretation, on the contrary, has no restriction of applicability whatsoever. Even if, at first sight, it might seem to lack a specific meaning, it actually gives rise to a very realistic approach, in light of the fact that an individual (or a society, a political organization, an insurance company, etc.) must somehow develop rules for making decisions under uncertainty for events that are not necessarily repeatable.

Note that when combinatorial probability can be applied, it coincides with the frequentist interpretation. Likewise, subjectivist and frequentist probability coincide in cases where the latter can be defined and is *known* (this last point is however a bit delicate and an explanation to the point would be too large a digression here).

A characteristic element of the subjectivist approach is that it aims to formalize the mechanisms by which the probability of an event varies with the state of information; it is this approach which has been essentially implied in the considerations developed here.

From a mathematical point of view, the theory of probability based on Kolmogorov's axioms and on *measure theory* has meanwhile shown a development of extensive proportions in the twentieth century. The evolution of this theory, stronghold of an extraordinary success that was obtained gradually, in particular in the theory of random processes, has somehow left aside the debate on the physical meaning of the term *probability*.

As in any mathematical theory, one can avoid worrying about the physical meaning of the objects one is studying, once the axioms which these have to obey are established.

It is a fact however, that in the study of probability and its applications there are situations that exceed the axiomatization put forth by Kolmogorov, or in which the axiomatization is revealed to be too restrictive. Substantially there are two aspects that have to be kept in mind (see [7]):

a) in some cases the assumption of countable additivity becomes too restrictive and the application of a weaker and more general theory of probability, which is based on the axiom of finite additivity only, may be more appropriate;
b) in other cases it is the very idea of being able to specify the *sample space* a priori that is not very realistic. In fact, it can happen that, while observing an experiment, an event that had not been a priori anticipated occurs, and this event reveals a whole range of other possibilities which one had not thought of (and which indeed have not taken place) but which however would then have to be considered as potentially observable.

The necessity of studying this second aspect can in particular emerge in situations where a subjectivist interpretation of probability appears natural, and may also occur in the analyses of coincidences that interests us here. Other phenomena, which when studied reveal an inadequacy of Kolmogorov's axiomatization (and which could have a certain relevance in the analysis of coincidences) are those studied in the field of "quantum probability" see e. g. [11]). On a methodological level it has to be underlined, however, that the previous points cannot in the least diminish the fundamental role that Kolmogorov's axiomatization has had in the theory of probability.

As was said, on the other hand, the cases considered under b) can have a certain relevance in the study of coincidences. The examples that we have outlined in the preceding section are however not of this type: we have only considered cases in which the identification of an appropriate "prepacked" sample space was not up for discussion. Nevertheless, an in-depth analysis of examples of the type mentioned above could also prove to be interesting.

## Bibliography

[1] R.K. Merton and E. Barber, *The Travels and Adventures of Serendipity: A Study in Sociological Semantics and the Sociology of Science*. Princeton University Press, Princeton, NJ, 2004.

[2] A. Koestler. *The Roots of Coincidence*. Vintage, New York, 1972.

[3] P. Odifreddi. *C'era una volta un paradosso*. Einaudi, 2001.

[4] P. Diaconis and F. Mosteller. Methods for studying coincidences. *Journal of the American Statistical Association*, 84:984–987, 1989.

[5] I. Peterson. *The Jungles of Randomness: A Mathematical Safari*. Wiley, New York, 1997.

[6] W. Feller. *An Introduction to Probability Theory and its Applications*, volume I. Wiley, 1968.

[7] B. de Finetti. *Teoria delle Probabilità*. Einaudi, 1970. English translation: *Theory of Probability. A critical introductory treatment*. Wiley, Vol. 1, 1974; Vol. 2, 1975.

[8] R. Arratia, L. Goldstein, and L. Gordon. Poisson approximation and the Chen–Stein method. *Statistical Science*, 5(4):403–434, 1990.

[9] A.N. Kolmogorov. *Grundbegriffe der Wahrscheinlichkeitsrechnung*. Erg. Math., Vol. 2, No. 3. Springer-Verlag, 1933. English translation: *Foundations of the Theory of Probability*. Chelsea Publishing Company, 1956.

[10] P. Billingsley. *Probability and Measure*. Wiley, 1979.

[11] S.P. Gudder. *Quantum Probability*. Academic Press, 1988.

# Mathematics and Design

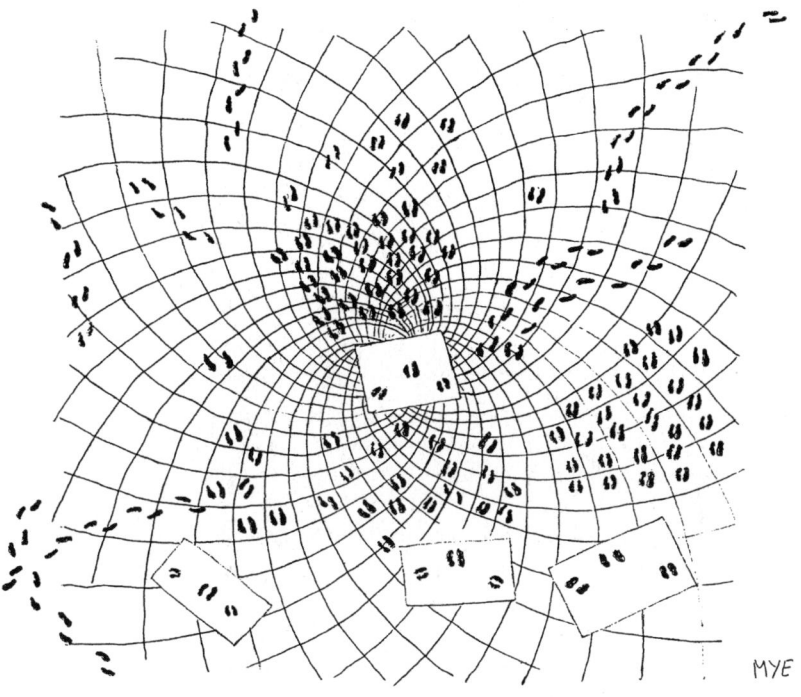

# The Square Fish*

Marco Campana

"The Square Fish" is an illustrated fable and, as all illustrated fables, in this case as well the images serve to make the narrative visible. This is, however, only a first impression. In reality, at the decisive moments, the images not only depict but "prove", if you will, the truth of the statements on a perceptive level as well as in the geometric sense. On the level of perception, in as much as the square fish sees the individual figures in a new set, coherent and different from how the other fish see them. In the geometric sense, because of the divisibility of the square into geometric shapes of the same kind. Thus it turns out that both the text and the images are absolutely necessary.

I have tried here to represent briefly the compositional course of the tale that describes the creative journey that I undertook in creating this fable. Everything started with the idea that, in a field of representation, the elements can be configured in different ways if we bring into play, for example, the perception of the edge of the representative field. Imagining and drawing a black sun and transversal cuts, which might be the entry portal to another dimension, I next took in consideration the edge of the page as if it were a drawing. That was, obviously, only one possibility (when we observe a painting or the page of a book, the relationship of shape formed by the edges to the content generally doesn't matter to us), but if we use it, it completely changes the whole and the visual meaning: the sun becomes an eye and the slits become the fish gills, and since the page was square, what I am looking at is a big square fish. Going from the large square to the small square, by subdivision, happened instantly.

* *Translated by Kim Williams*

## The Square Fish

**Scene 1**  In the calm and transparent waters of a faraway ocean arrived, when it was time, the current from the south, and it passed lightly through the fish eggs among the algae.

**Scene 2**  So were born that year's new fish and right away they grouped themselves into families. Only one didn't know where to go, and all the others asked him, "What kind of fish are you?"

**Scene 3**  The little fish didn't answer, and all by himself he went along an ocean path until he came to a large rock. He swam to the top, and there he stopped.

**Scene 4**  There he began to watch the world, and day after day he stayed on that rock, sadder and sadder, and lonelier and lonelier, while everything around him changed and grew. Eels swam by him. One of them saw him, and went over.

**Scene 5**  The lonely fish asked, "Do you know me?" The eel stretched out to his full length, then he folded himself into four equal parts until his head touched his tail, and said, "See? You are a square fish." Then he stretched himself out again and swam away.

**Scene 6**  A little while later, in the middle of the day, the water darkened and to the snails who were fleeing, hanging on to the algae, the fish asked, "Why are you going away?" "Strange things are happening in our ocean – they answered – you should run away too" ...

**Scene 7**  But the square fish went to meet the other fish, and found them all huddled together, afraid. "What are those slits that open and close and move all of the water near them?" he asked one of them. "It's a door to the ocean of the dead," answered one ...

**Scene 8**  "And that circle high above that we can sometimes see is its sun, always black", said another. But the square fish said, "The thing that opens and closes is its breathing, and that dark circle is its eye. Now look at the whole together with the edges: it's a very big fish, a square fish like me" ...

**Scene 9**  Then he moved a little bit closer and asked, "Are you a bad fish or a good fish?" The big fish looked at him and smiled. "You're a good fish – said the little fish – but too big" ...

**Scene 10**  So the big fish transformed himself and the whole ocean was filled with eyes.

**Scene 11**  Then, one after the other, they all took the form of little square fishes.

**Scene 12**  And they spread through all the oceans of the world ...

**Scene 13**  But some stayed, and from that day on the square fish lived with them, happily ever after, in the big fish family ...

**Fig. 1.** The Square Fish – illustrations.
(see the section in colour)

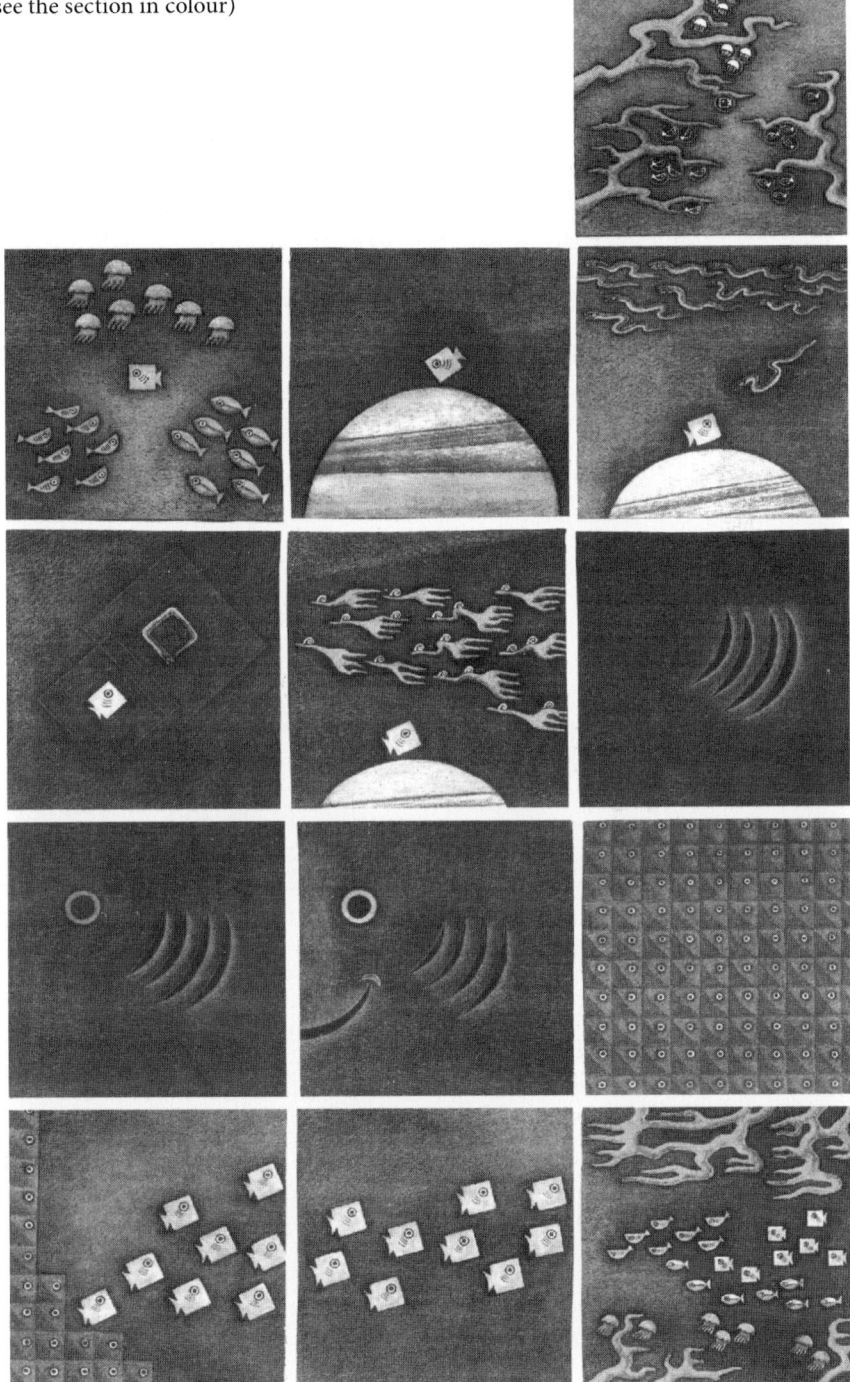

123

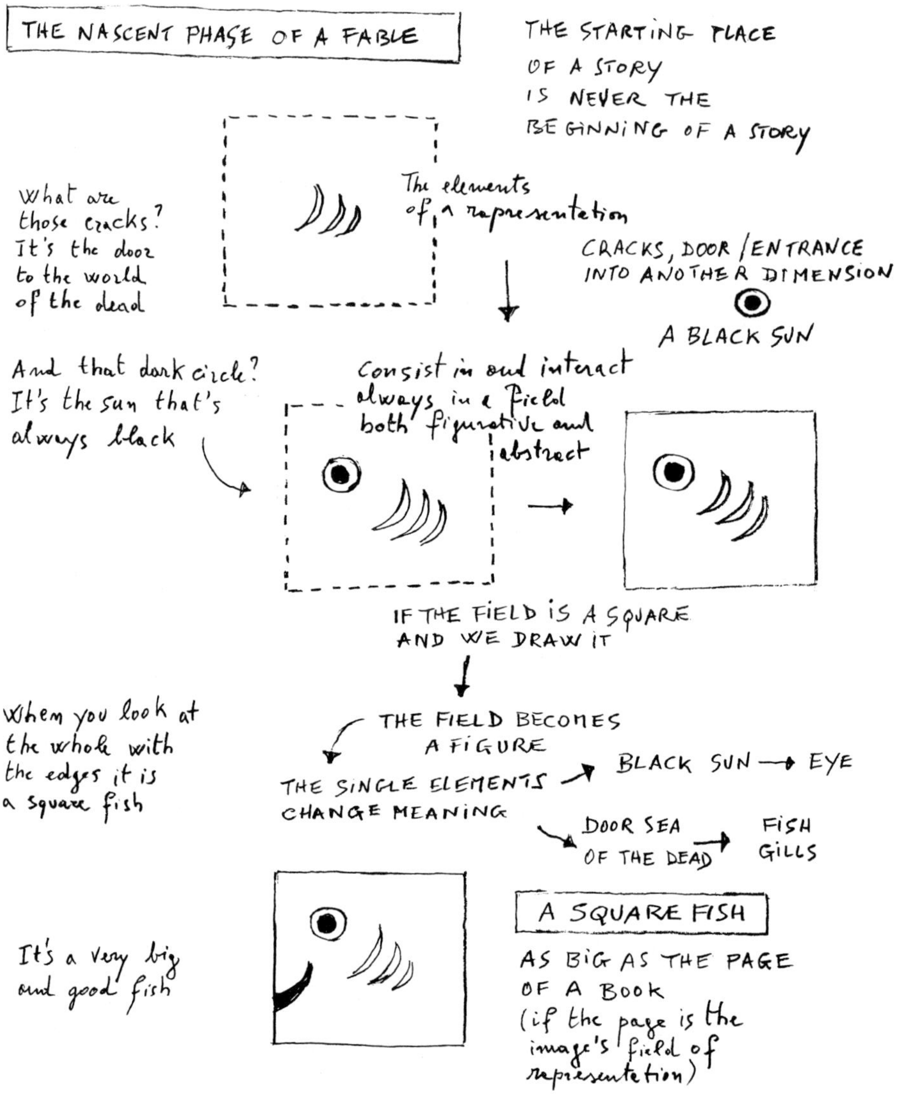
THE NASCENT PHASE OF A FABLE
THE STARTING PLACE OF A STORY IS NEVER THE BEGINNING OF A STORY
What are those cracks? It's the door to the world of the dead
The elements of a representation
CRACKS, DOOR / ENTRANCE INTO ANOTHER DIMENSION
A BLACK SUN
And that dark circle? It's the sun that's always black
Consist in and interact always in a field both figurative and abstract
IF THE FIELD IS A SQUARE AND WE DRAW IT
When you look at the whole with the edges it is a square fish
THE FIELD BECOMES A FIGURE
THE SINGLE ELEMENTS CHANGE MEANING
BLACK SUN → EYE
DOOR SEA OF THE DEAD → FISH GILLS
A SQUARE FISH
It's a very big and good fish
AS BIG AS THE PAGE OF A BOOK (if the page is the image's field of representation)

**Fig. 2.** The creative process of the square fish fable, scheme 1

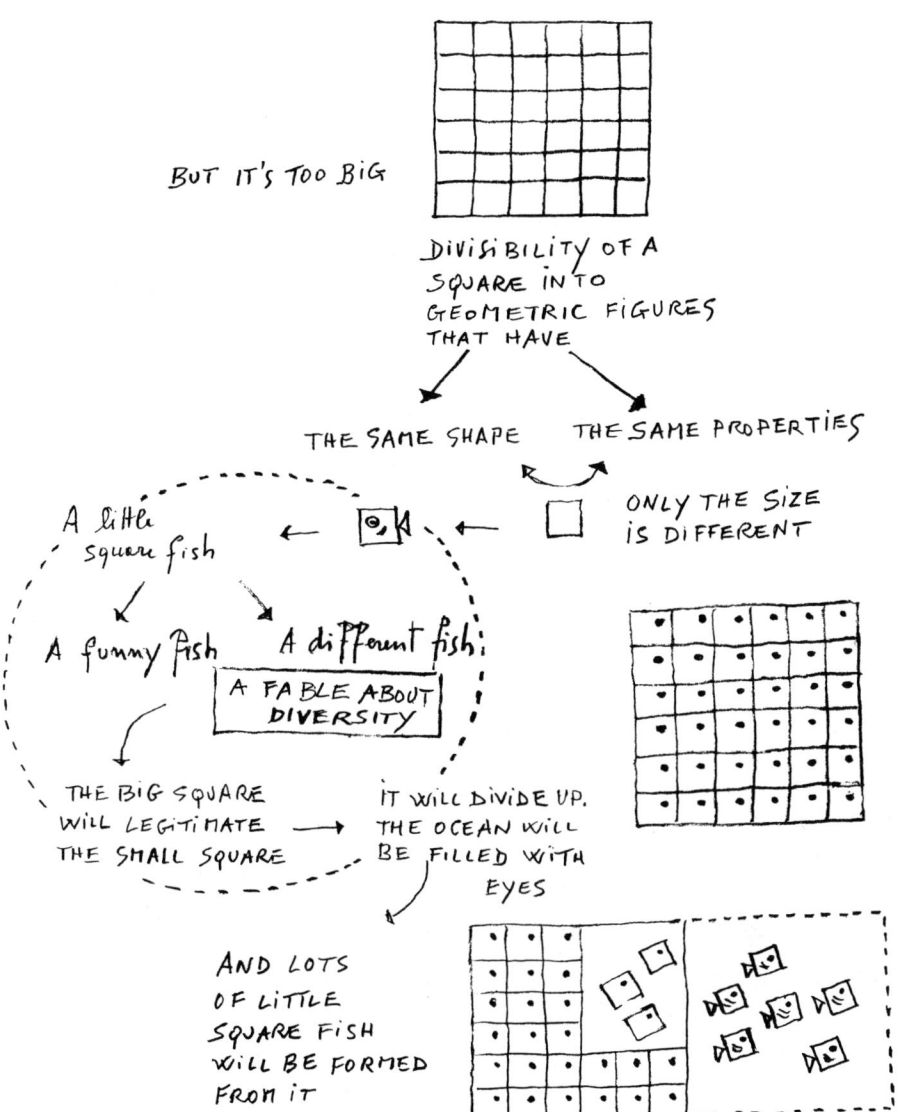

**Fig. 3.** The creative process of the square fish fable, scheme 2

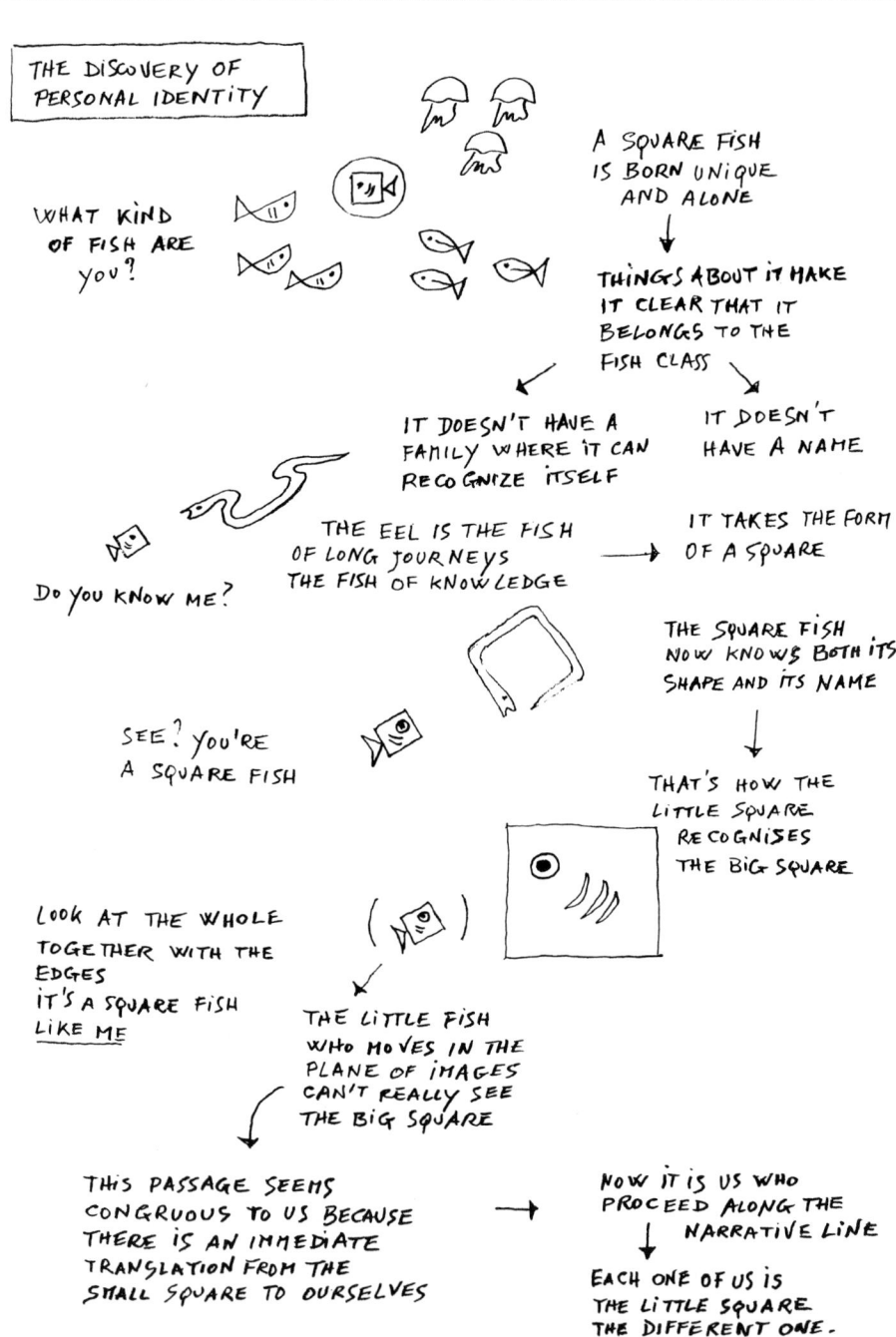

**Fig. 4.** The creative process of the square fish fable, scheme 3

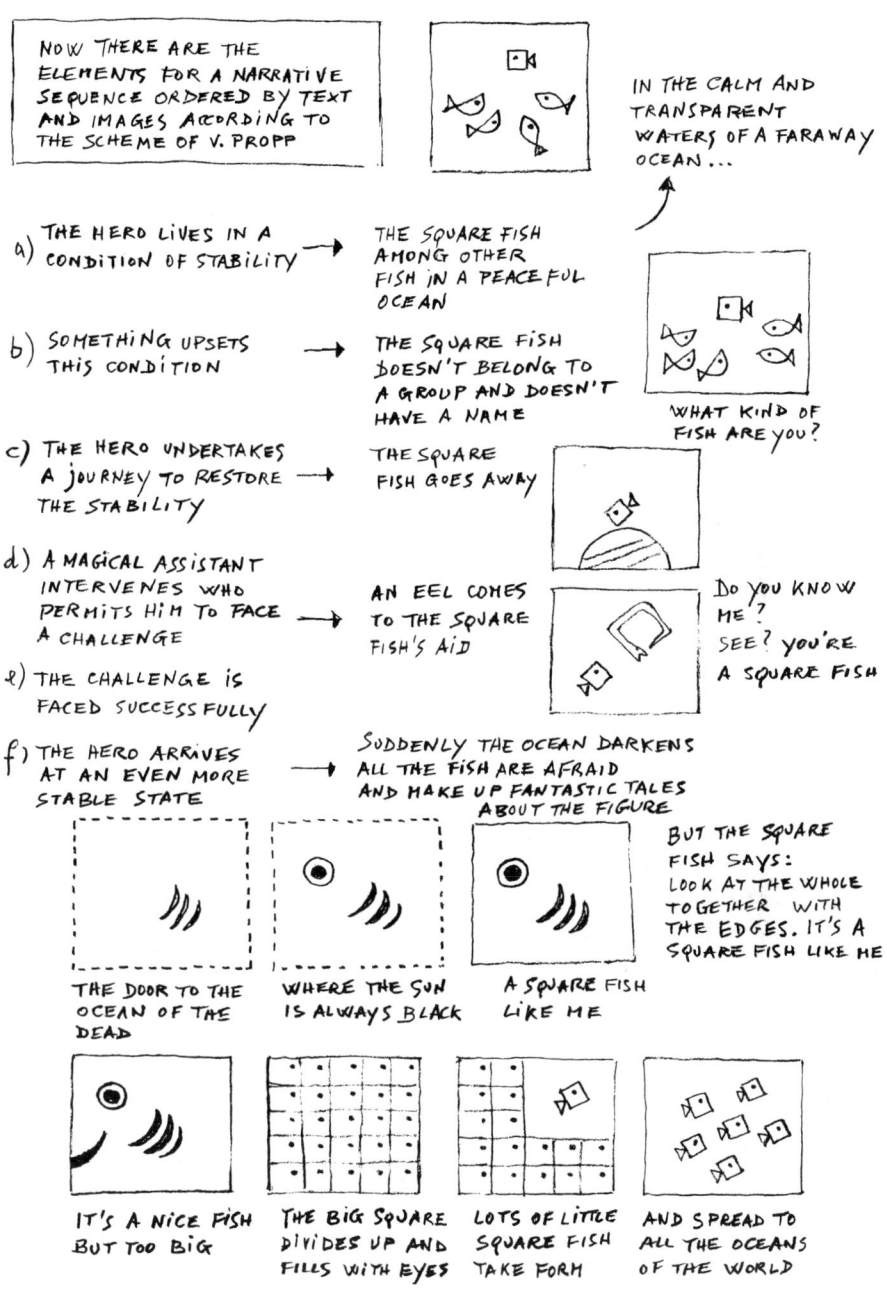

NOW THERE ARE THE ELEMENTS FOR A NARRATIVE SEQUENCE ORDERED BY TEXT AND IMAGES ACCORDING TO THE SCHEME OF V. PROPP

IN THE CALM AND TRANSPARENT WATERS OF A FARAWAY OCEAN...

a) THE HERO LIVES IN A CONDITION OF STABILITY → THE SQUARE FISH AMONG OTHER FISH IN A PEACEFUL OCEAN

b) SOMETHING UPSETS THIS CONDITION → THE SQUARE FISH DOESN'T BELONG TO A GROUP AND DOESN'T HAVE A NAME

WHAT KIND OF FISH ARE YOU?

c) THE HERO UNDERTAKES A JOURNEY TO RESTORE THE STABILITY → THE SQUARE FISH GOES AWAY

d) A MAGICAL ASSISTANT INTERVENES WHO PERMITS HIM TO FACE A CHALLENGE → AN EEL COMES TO THE SQUARE FISH'S AID

DO YOU KNOW ME? SEE? YOU'RE A SQUARE FISH

e) THE CHALLENGE IS FACED SUCCESSFULLY

f) THE HERO ARRIVES AT AN EVEN MORE STABLE STATE → SUDDENLY THE OCEAN DARKENS ALL THE FISH ARE AFRAID AND MAKE UP FANTASTIC TALES ABOUT THE FIGURE

BUT THE SQUARE FISH SAYS: LOOK AT THE WHOLE TOGETHER WITH THE EDGES. IT'S A SQUARE FISH LIKE ME

THE DOOR TO THE OCEAN OF THE DEAD

WHERE THE SUN IS ALWAYS BLACK

A SQUARE FISH LIKE ME

IT'S A NICE FISH BUT TOO BIG

THE BIG SQUARE DIVIDES UP AND FILLS WITH EYES

LOTS OF LITTLE SQUARE FISH TAKE FORM

AND SPREAD TO ALL THE OCEANS OF THE WORLD

**Fig. 5.** The creative process of the square fish fable, scheme 4

127

# The Square: Homage to Bruno Munari*

Michele Emmer

> *The square is not a subconscious form.*
> *It is the creation of intuitive reason.*
> *The face of the new art!*
> *The first step of pure creation in art.*
> K. Malevich

## Malevich and the "Black Square"

In December 1913, in the theatre of the Luna Park on Ulitza Oficerskaja, in St. Petersburg, *Victory Over the Sun* was staged for the first time. The design of the performance had been decided on in a meeting in Finland, from 18 to 20 July of that same year, between the poet Aleksja Krucenych, the composer Mihail Matjusin, and the painter Kaismir Malevich. The script was written by Krucenych, the music composed by Matjusin, the costumes and set design realised by Malevich. Of that performance, apart from the script, we have very few images today.

Benedikt Livsic watched the performance, and wrote of it:

Malevich's design showed his novelty and originality first of all in the use of light as a principal element creating form [...] Within the confines of the scenic box a pictorial stereometry was born, determining a rigorous system of volumes, so as to reduce to a minimum the elements of chance that the movement of human figures caused from the exterior. The figures themselves were sliced by the razors of light: from time to time they lost hands, feet, heads, because for Malevich these were only geometric entities, which could not only be severed from their respective components, but also dissolved completely in the pictorial space. [1]

---

* *Translated by Kim Williams*

This performance, never again repeated, will remain a significant moment in the art of the twentieth century; a reconstruction of it, based on the text and the few staging and set design materials that remain, was realised by the County Museum of Art in Los Angeles. I was able to see this performance in a video, obviously in English, during the exhibition of Malevich's works in 2005 at the Museo del Corso in Rome.

The evening in St. Petersburg was not an indisputable success. Some loved the performance, while others rejected it altogether. The performance was Cubist-Futurist but, as Josef Kiblickij wrote in the catalogue of the Roman exhibition:

> Only a little later would Malevich understand the great importance of that moment, and that precisely then, in December 1913, he was testing the land in search of a new way: an escape from Futurism. Examining the imperceptible threshold of the backdrop with the black square, directly lacerated on stage, Malevich dived into a world and a dimension that were entirely new: the future. He realised only later that he had reached a significant milestone, when he reinterpreted and re-elaborated the costume designs, which contained some compositional schemes of what would become Suprematism. [1]

As Jean-Hubert Martin has written of the set design:

> ... the only reality is the abstract form. Malevich's set design has made very evident the importance, in treating abstract form, of the internal logic of the artistic form, conceived first of all as its composition. From the first scene, and then in the second and third, there appeared schematic indications of a Suprematist square. In the fifth scene the backdrop consisted of a black square, which stood out against the white surface. [2]

It would be just two years later that Malevich for the first time showed in an exhibition *The Black Square* (Fig. 1).

Malevich wrote to Matjusin, composer of the music for the performance:

> I would be most grateful if you could publish one of my sketches for the set design. This drawing will be of enormous importance for painting. It represents a black square, the embryo of all possibilities which, in their development, acquire a surprising strength. It is the progenitor of the cube and sphere and its dissociation produces a cultural contribution that is fundamental for art. [2]

In December 1915 the artists of Moscow and St. Petersburg put on an exhibition that was described by the symbolic-numeric denomination "0–10" (the complete title was *Last Futuristic Exhibition of Paintings*); on this occasion Malevich published a pamphlet in which the word "suprematism" appears for the first time, and *Black Square* stood out among the works of Malevich that were shown. As Kiblickij wrote, "a new religion of representation was born. It began again counting from zero". As though recalling the title of the exhibition, Malevich stated the concept of the *zero point of the form*: "I am transfigured in the zero point of the form".

In the catalogue of the exhibition Malevich pointed out that in giving a title to these paintings, he didn't want to suggest which form should be looked for, but

**Fig. 1.** K. Malevich, *Black Square*

rather to indicate that the true forms are often considered as points of departure for indefinite pictorial masses, which have given life to a pictorial painting without any reference to nature.

Purity of form, their apparent clarity of perception, a perception of "physically palpable" space.

*Victory Over the Sun* was the title of the opera, because the sun is the symbolic expression of the values dear to the poetic tradition of the past, which had become by then banal stereotypes. The sun, symbol of the old aesthetic, and of the "mean mathematical rationality" as well.

> *We have eradicated the sun with its fresh roots*
> *fat, stinking of arithmetic*
> *here it is, look*
> "Victory Over the Sun", scene 4, part 1

But in mathematics there is much more than the "roots that stink of mathematics". In the work *Victory Over the Sun*, and in particular in the second *agimento* (part), the laws of physics and social conventions are abolished, the spatio-temporal coordinates are unhinged, logical-causal laws are no longer valid, and consequentially, neither are the rules of grammar that express them: "time runs backwards, cause can precede effect, the force of gravity no longer works". It is the world in reverse. A great liberty is taken also with the consideration of space. Malevich comes to define himself as "the president of space".

From what are derived these new images of the world in reverse, in which the spatio-temporal coordinates of reality are of no use, and neither are the logical models that regulate our interpretation of it? As in all attempts to retrace the cultural threads that interweave and produce an influence in the most diverse sectors and environments, here we are not dealing with an exclusive link that has a direct influence, whether of a mathematical nature or not. Besides the obvious echoes of traditions of folklore and Carnival, the Russian futurists dip into some texts of a pseudo-scientific or paraphilosophical nature: in short, Malevich's black square has a history.

Linda Henderson writes:

> A-logic was not the final answer to the problem of representing the fourth dimension in art, as happened in literary style. If the capacity of the writing of Krucenych to evoke a new form of awareness reached its acme in *Victory over the Sun*, at the same time Malevich began to experiment with the geometric shapes with which he would ultimate obtain an expression of the fourth dimension in the Suprematism of 1915. In December of 1913 many of these geometric experiments do not seem to have any specific connection to the fourth dimension. However, some groups of drawings created for the opera could be associated with the fourth dimension. In both the background that represents the house in the second act of *Victory* and in the 1913 drawing *Strumento musicale/lampada* (Musical Instrument/Lamp) Malevich seems to have incorporated the popular image of the four-dimensional hypercube in this drawing [3].

Square, cube, hypercube. Why the beginning of the twentieth century? Does that black square have a history? And the four-dimensional cube?

## The Square Protagonist of "Flatland"

I (the square): What therefore more easy than now to take his servant on a second journey into the blessed region of the Fourth Dimension, where I shall look down with him once more upon this land of Three Dimensions, and see the inside of every three-dimensioned house, the secrets of the solid earth, the treasures of the mines in Spaceland, and the intestines of every solid living creature, even of the noble and adorable Spheres. In One Dimension, did not a moving Point produce a Line with two terminal points?

Sphere: But where is this land of Four Dimensions?

I: I know not: but doubtless my Teacher knows.

Sphere: Not I. There is no such land. The very idea of it is utterly inconceivable.

I: Not inconceivable, my Lord, to me, and therefore still less inconceivable to my Master. Nay, I despair not that, even here, in this region of Three Dimensions, your Lordship's art may make the Fourth Dimension visible to me; just as in the Land of Two Dimensions my Teacher's skill would fain have opened the eyes of his blind servant to the invisible presence of a Third Dimension, though I saw it not.

The characters of this dialogue are two regular geometric figures; the first, the main character of the story, the narrator I, is the Square, inhabitant of Flatland; the second character, treated with great deference by the other, is the Divine Sphere of the realm of three dimensions. We are in Flatland, the name of the country where the square lives is a flat world, a world of two dimensions. The square himself suggests to us an idea of his world (Fig. 2):

> Imagine a vast sheet of paper on which straight Lines, Triangles, Squares, Pentagons, Hexagons, and other [geometric] figures … move freely about, on or in the surface, but without the power of rising above or sinking below it, very much like shadows – only hard and with luminous edges …

Naturally, in Flatland "it is impossible that there should be anything of what you call a 'solid' kind". The inhabitants of Flatland cannot even imagine the existence of a three-dimensional object, since in order to measure a three-dimensional object there needs to be a three-dimensional unit of measure; from their point of view, it should be said, there exist only luminous lines that represent themselves, that is, the inhabitants, the houses, the trees of Flatland. Let us suppose that we lay a triangle on a plane, a table, and that we imagine looking at it with our eyes at the table's edge; we would see only a line, a segment. Conclusion: a square in Flatland could not have any idea at all of a Sphere; the merely supposition of the existence of a three-dimensional figure causes such a disruption of the country's tranquillity that anyone daring to think in this way is immediately arrested. Not only can no inhabitant have any idea of a Sphere, but if, as in our case, a Sphere were to come down from space to visit Flatland, no one would even recognise it because the only thing visible to two-dimensional eyes would be a line, which would represent the section between the plane where the inhabitants of Flatland live and the Sphere. Unless … a square were to meet a Sphere who makes him exit the plane and rise into Spaceland. The

133

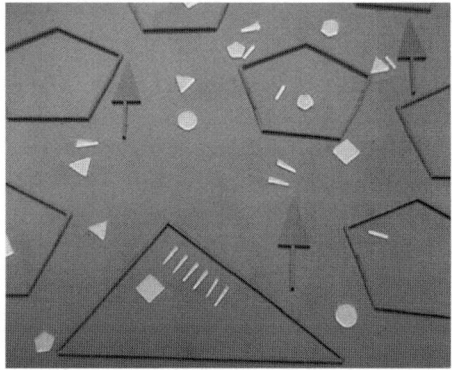

**Fig. 2.** From the film *Flatlandia* by Michele Emmer

Square then experiences what we would experience if someone were to pull us up into four-dimensional space:

> An unspeakable horror seized me. There was a darkness; then a dizzy, sickening sensation of sight that was not like seeing; I saw a Line that was no Line; Space that was not Space: I was myself, and not myself [...] "Either this is madness or it is Hell."

But anguish soon gives way to wonder:

> ... a new world! There stood before me, visibly incorporate, all that I had before inferred, conjectured, dreamed, of perfect Circular beauty. What seemed the centre of the Stranger's form lay open to my view: yet I could see no heart, nor lungs, nor arteries, only a beautiful harmonious Something – for which I had no words; but you, my Readers in Spaceland, would call it the surface of the Sphere.

The encounter between the Divine Sphere and the Square is the central event in a very famous book, particularly in the English-speaking world, a book whose complete title is *Flatland: a Romance of Many Dimensions*, the first edition of which was published anonymously in 1884 [4]; the author was an English theologist, a Shakespeare scholar and mathematics teacher named Edwin Abbott (1838–1926). The first edition came out without the author's name because Abbot was not very sure that it would be very good for his reputation if it were known that he, a scholar of the Bible and of Shakespeare, had written this kind of book.

A square the leading character of a story, a novel. A square who dreams of seeing what no one before him had seen, a divine cube in four-dimensions. A story that would have a great influence in the worlds of European and Russian literature and art. That square would become one of the protagonists of modern art.

## Art in the Twentieth Century

In 1949 [5] Max Bill (Fig. 3) wrote that the point of departure for a new concept of art in which mathematics would have an important role was probably due to Kandinsky, and that it was then Mondrian who more than any other distanced himself from the traditional concept of art. A mathematical approach to art, a new mathematical approach to art in which, paradoxically, the fact of having had information about the crisis of Euclidean space, of the birth of a four-dimensional geometry, of non-Euclidean geometries and of the mysterious and sometimes mystical fascination born of these new ideas, pushed some artists to a rediscovery of the essentialness and objectivity of geometric shapes, of elementary geometrical shapes. As Jean Clair writes:

> Everything happens as if the theory of the fourth dimension, under the apparent uniqueness of Euclidean space, provokes a "whirl" of different spaces, and

finally through the reflections of Moebius, Klein and Poincaré, the emergence of topology as a sciences, pushes the painter to consider the *failure* of three-dimensional space and of its perspective projection, leading the artists to consider nothing other than the properties of the space of their own discourse, in other words, of the plane space of the canvas; the necessarily flat space, without thickness, of purely formal effects. At the moment in which mathematical analysis led to topology as a science, painting could not help but lead to *tropology* as art. What Marcel Duchamp would call a "*pictorial nominalism*" [6].

No more, as Cezanne also declared, geometric shapes seen in perspective, but the imposition of a 90 ° rotation of the painting, substituting a kind of planimetry for perspective, that leads it to merge with the ground plane. As Poincaré said, the painting is a section, a cut, a section that a two-dimensional tool – that is, the painting – performs on a phenomenon that has several dimensions. As Clair writes, "The painting is the plane section of the orthogonal projection of the phenomenon being considered".

In short, the world where the square of Flatland lives becomes the space of privilege of art and the forms that live there are those that live in Flatland. This then is how to understand the influence of Abbot's book and those of Hinton and Wells that followed, as well as the famous *Black Square on a White Background* by Malevich; Clair emphasises that it will no longer be

> … the manifestation of some irrational heroism, as one part of art criticism would still have it today, but rather, in this new logic established by topology, its maximum tropism, that is, suprematist, which sets itself as the ultimate synecdoche of the chequerboard seen in cavalier perspective. An effect of the style that exceeds representation, the starting point for founding the perspective of the art of the future is the new square. […] In place of the representation of appearances is substituted the manifestation of the essence of things [6].

The chequerboard is no longer seen in perspective, squares are no longer seen in perspective, but represented as sections of objects with several dimensions. And the break came after 1910, for example, with Juan Gris who, re-using the motif of the playing card (rectangles), rigorously represents them as identical, superimposable, "motif, but above all symbol", underlines Clair "of a perfectly flat space that the square has become" (Fig. 4).

From this, then, the use of geometric shapes, of squares, becomes obvious. It is one of the characteristics of the new relationships between mathematics and art at the beginning of the twentieth century, as two art historians, Lucy Adelman and Michael Compton, write in an essay entitled "Mathematics in Early Abstract Art" [7], regarding the artistic avant-garde:

> Even though painting and mathematics are two very different disciplines that are often considered totally opposite, between the two of them there are many points in common.

**Fig. 3.** Max Bill in his Zurich studio, from the film "Ars Combinatoria" by Michele Emmer

**Fig. 4.** M. Bill, "Variations", from the film "Ars Combinatoria" by Michele Emmer

136

The authors note how at the beginning of the twentieth century there is a reconciliation between art and mathematics which will turn out to be fruitful to the former. Different levels of relationships can be distinguished between mathematics and art, levels that the two authors identify for the sake of convenience, given that often different levels are found together in the works of an artist, or even within a single work. We are talking about the beginnings of abstract art.

> Above all there was an widespread interest in non-Euclidean and/or $n$-dimensional geometries [...] Secondly, the period marked the undoing of perspective and its substitution with different, less systematic canons. Third, the artists made use of numerical proportions or grids that were, like geometric shapes, associated with the idea of reducing art to its specific elements. Fourth, there appear in painting elements that are taken from mathematical texts [...] Finally, simple geometric figures came to be associated with machines and their products and in this way with *progress* and *modernity*. [...] The regular shapes in two and three dimensions and the proportions were studied by more ancient civilisations and were believed to embody special properties of beauty, of universality, and even of divinity.

Thus we have the theme of chequers, both for the square aspect to be represented without perspective and for the aspect of the grid, which had become of great interest to artists. One of the chapters of the book that Jean Clair dedicated to Marcel

Duchamp deals specifically with chequers [8]. The flat, two-dimensional surface of the chequerboard merges, becomes one with the surface of the canvas; the canvas, writes Clair, citing Alberti's definition of perspective, ceases to be "the open window where I see what will be painted there", becoming instead a flat space, without depth, placed vertically, a true grid onto which to place various elements, as the pieces are placed on the chequerboard.

Mondrian provides an extreme example when he makes of the painting an abstract orthogonal grid, then realising in 1917 a distribution of planes on the canvas in a seemingly random order, and then making the surface of the painting a genuine chequerboard in 1919.

The square was an essential shape for art at the turn of the nineteenth century; a sort of archetypal form that, even in its simplicity, seems to provide the painting with certainty, with a solid basis. The new geometries, the new spatial freedoms, would lead to a great interest in one of the primary geometrical shapes described in Euclid's *Elements*, two thousand years earlier.

## The Square and Design

Mathematicians are creators of form, discoverers even, when one thinks that mathematical ideas such as the Platonic ideals existed before our thoughts about them did. In our days, with the great dissemination of computer graphics it is easy enough for anyone to verify that that's the way things are. Suffice it to think of the great graphic revolution caused by fractal geometry. Culture and design are profoundly touched by the new forms that mathematicians create. A rather sensational example is the use of topology, from the Moebius strip to the Klein bottle (one has only to recall the influence on contemporary architecture).

Bruno Munari has written about topology and has used it, for example, in his book *Arte come mestiere* [9], in which a chapter is dedicated to the Moebius strip to topology. One form that Munari has used quite a lot is, in fact, the square.

In 1978 in the book series entitled *Quaderni di design*, Munari published the volume entitled *La scoperta del quadrato: più di trecento casi in cui tutto ciò ha una ragione per essere quadrato* (The discovery of the square: more than three-hundred cases in which everything is square for a reason) [10]. In the introduction, he writes (Fig. 5):

> In the most ancient writings and in rock carvings, the square stands for the idea of enclosure, house, town, field. It is a shape that is rather rare in nature, where it is found in the cubic pyrite of the island of Elba, in some crystals, and in some structures that are revealed by an electron microscope. In the architecture of various peoples we find from the most ancient times buildings with a square plan, especially those for collective, religious and defensive purposes; many castles had square plans. A square grid governed the plan of many cities, still today many architects build their buildings on square plans.

137

The book contains the poem by Carlo Bellolli:

| | |
|---|---|
| *Il campo* | *The square* |
| *quadrato* | *field* |
| *la piazza* | *the square* |
| *quadrata* | *square* |
| *la città* | *the square* |
| *quadrata* | *city* |
| *la prigione* | *the square* |
| *quadrata* | *prison* |
| *la tomba* | *the square* |
| *quadrata* | *tomb* |
| *la tenda* | *the square* |
| *quadrata* | *tent* |
| *la pelle* | *the square* |
| *quadrata* | *skin* |
| *la pupilla* | *the square* |
| *quadrata* | *pupil* |
| *il quadrato* | *the square* |
| *è* | *is* |
| | |
| *la società* | *society* |

*Delle poesie della geometria elementare*, Basel (1959) (translation by Kim Williams)

138

**Fig. 5.** *La scoperta del quadrato* by Bruno Munari

**Fig. 6.** Bruno Munari in his studio in Milan, from the film "L'avventura del quadrato" by Michele Emmer

Some years later, in 1978, Gillo Dorfles wrote in *Viaggio nel quadrato* (Journey into the square) [11]:

> Even a simple square, even a rectangle, can vibrate, can be vital, can overcome its essentially geometrical nature and become loaded with impulses and tensions that translate it into an element that is autonomous and – let us even dare to say the dangerous word – *organic* [...] A precise code, clearly defined, closed to any formal adventure, to any decorative whim. And, precisely for this reason, even more arduous to follow, more severe, more circumscribed. And yet, within this meagre compositional possibility – where the square and rectangle dominate, and rarely there appears the outline of a triangle – we immediately note the presence of a enlivening element; whether it is a very peculiar quality of colour, or a certain new and unexpected "openings" of the form.

139

And Munari, again in the introduction of *La scoperta del quadrato*, speaking of modern art, added:

> In the field of visual art the square is the spatial module of which or with which, visual operators, researchers, and experimenters find various ways of structuring their works. These measures, including the famous golden section, derive from operations on the square, modifying it on the basis of precise rules of decomposition and recomposition, derived from the logical subdivision of its own dimensions, whether the space within the square or by carrying some of its internal measurements to the exterior, or subdividing it with the use of a compass or straightedge. In the field of graphics the square helps to structure many works, from manufacturers' logos, to symbols, to signals.

The book is full of examples, three hundred of them, as the title makes obvious, from the squares of Joseph Albers to Mayan architecture, from Max Bill to the Sforza Castle in Milan, from Franco Grignani to Klee to Le Corbusier, to El Lissitsky, from Morellet to Malevich. Nor is the hypercube neglected. With some works of Munari himself, such as *Concavo-convesso* of 1948. Take a square of wire mesh and curve it so that the corners touch at pre-determined points (in the film "L'avventura del quadrato" [12], filmed during the 1980s in his Milan studio, Munari suggested that you could even do this with the dough that tortellini are made from). The nets thus folded can be hung from the ceiling and, moving along the walls, produce moiré effects that continuously transform (Figs. 7 and 8).

An elementary figure, the square, but complex, not simple.

Munari also considered the case of the Peano curve:

> Peano curve. The mathematicians once held that every curve had to have tangents, because this property is intuitively evident. But in 1890 the famous mathematician Peano proved that there can exist a type of curve that cannot have tangents at any point. Since a circle can also be considered as a polygon with an infinite number of sides, then in the same way a broken line can be considered a curve. When this curve is drawn in a square space subdivided into many smaller squares, and following a progressive course so that it becomes increasingly dense within that same space, Peano proved that at its maximum density the curve would have filled the entire area of the square space. In this case there wouldn't be any space for the tangents.

The square is a shape that has remained unscathed by humanity's artistic adventures because it is "without style", as Munari says in the film "L'avventura del quadrato". And among the artists cited in the book there is, of course, El Lissitsky.

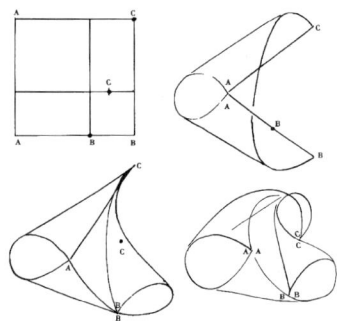

**Fig. 7.** Munari, Project for "Concavo-convesso"

**Fig. 8.** "Concavo-convesso" from the film "L'avventura del quadrato" by Michele Emmer

# El Lissitsky

Malevich considered the square the purest shape, contrary to Kandinsky who saw purity in the shape of the circle. As Adelman and Compton [7] write, El Lissitsky understood clearly that the declaration of the axiomatic idea of geometry (as set forth by Lobachevsky, Gauss, Riemann, and Bolyai) was at the root of the challenge to the idea that Euclidean geometry was the only "true" geometry.

> Even if Lissitsky had used traditional forms, right angles, triangles, circles, it is clear that he considered them as variations in tone, in color, in the composition, in rotation, part of a group of specific tools with which to work in the two-dimensional dimension. He combined them with more sophisticated curves such as ellipses, parabolas and hyperbolas that, even though two-dimensional, could represent various kinds of space.

El Lissitsky was another of the artists who looked to the power of geometry to bring order to the world. He worked with Malevich for three years on an essay entitled "A and Pangeometry", where A stands for "Art" and refers to the title of the book by Lobachevsky. Lissitsky believed that the Euclidean concept of an immobile space had been destroyed by Lobachevsky, Gauss and Riemann. "He identified mathematics with the rational, with the conscious side and with the means of the artist", write Adelman and Compton. Jean Clair writes that Lissitsky performed a kind of synthesis of the speculations about the fourth dimension that had begun to develop beginning in 1880.

141

> He presented them in the form of a diagram [...] on one side was drawn the classic pyramid of visual rays with the eye of the observer as its vertex and the object being considered as its base. On the other he drew the new system of projection: a system such that the rays of the pyramid have become parallel, in other words, where the eye has been transposed to infinity [8].

In this new space forms ceased to be apparitions that vary according to the position of the subject – on the other hand they would no longer be subject to elongation or foreshortening, but would always remain identical to themselves in all locations.

The first painting entitled *Prouns* (which stands for *proekt utverzhdeniya novogo*, "project for the affirmation of the new") dates from 1919. In the magazine *De Stijl* in 1922 Lissitsky wrote that,

> *Prouns* moves towards the construction of space, divides it into elements in all dimensions and constructs a new image, unitary even if of many faces, of nature.

Lissitsky arrives at the conclusion that, as he writes in "A and Pangeometry":

> Multidimensional spaces that exist mathematically cannot be conceived, cannot be represented, and therefore cannot be materialised.

**Fig. 9.** El Lissitsky, *Red Square*, from the film "L'avventura del quadrato" by Michele Emmer

In 1922 Lissitsky published in *De Stijl* the story of a "red square that defeats the black square", restoring order from chaos. A revolutionary square, like the square of Flatland some decades earlier who had been put into prison for that reason. A Bolshevik square who defeated dark, black evil. The apotheosis of the square that, in the key of the International, changed the world (Fig. 9).

This shape, we want to say, so ancient, so familiar, so banal. A shape that would become the emblem of the new geometry of the art of the twentieth century. Or of the end of art, as the title of the book that Jean Clair dedicated to Duchamp says.

In the preface of Munari's book is a saying by Maxim Gorky:

> An artist is one who is able to exploit his own subjective impressions, knows how to discover an objective general meaning and express it convincingly.

## Bibliography

[1] J. Kiiblickij (2005) Sulla questione del quadrato nero nell'opera *Vittoria sul sole*, in: C. Beltramo, C. Zevi (eds.), *Kazimir Malevich, oltre la figurazione e oltre l'astrazione*, Artificio Skira, Firenze, pp. 145–149.

[2] J.-H. Martin (1978) Kasimir Malevich: fonder une ère nouvelle, in: P. Hulten, J.-H. Martin (eds.), *Malevitch*, centre George Pompidou, Paris, pp. 9–20.

[3] L.D. Henderson (1983) *The Fourth Dimension and Non-Euclidean Geometry in Modern Art*, Princeton University Press, Princeton, p. 277.

[4] E.A. Abbott (1884) *Flatland: a Romance in Many Dimensions*, Seeley and Co., London.

[5] M. Bill (1993) The Mathematical Way of Thinking in the Visual Art of Our Time, in: M. Emmer (ed.), *The Visual Mind*, MIT Press, pp. 5–9. This article was published the first time in German in 1949 in *Werk* 3, and was republished several times with the title "The Mathematical Approach in Contemporary Art". In the volume cited here it was revised by the author with some slight corrections.

[6] J. Clair (2000) *Sur Marcel Duchamp et la fin de l'art*, Gallimard, Paris, p. 132.

[7] L. Adelman, M. Compton (1980) Mathematics in Early Abstract Art, *Towards a New Art. Essays on the Background to Abstract Art 1910-20*, The Tate Gallery, London, pp. 64–89.

[8] J. Clair (2000) *L'échiquier à trois dimensions*, in *Sur Marcel Duchamp et la fin de l'art*, Gallimard, Paris, pp. 111–133.

 [9]  B. Munari (1975) *Arte come mestiere*, Laterza, Bari.
[10]  B. Munari (1978) *La scoperta del quadrato*, Quaderni di design, Zanichelli, Bologna.
[11]  G. Dorfles (1978) *M. L. de Romans viaggio nel quadrato*, Edizioni Bora, Bologna.
[12]  M. Emmer (1984) *L'avventura del quadrato*, film, video, DVD, 25 minutes, in the series *Art and mathematics*, © M. Emmer, Rome.

143

# Mathematics and Cartoons

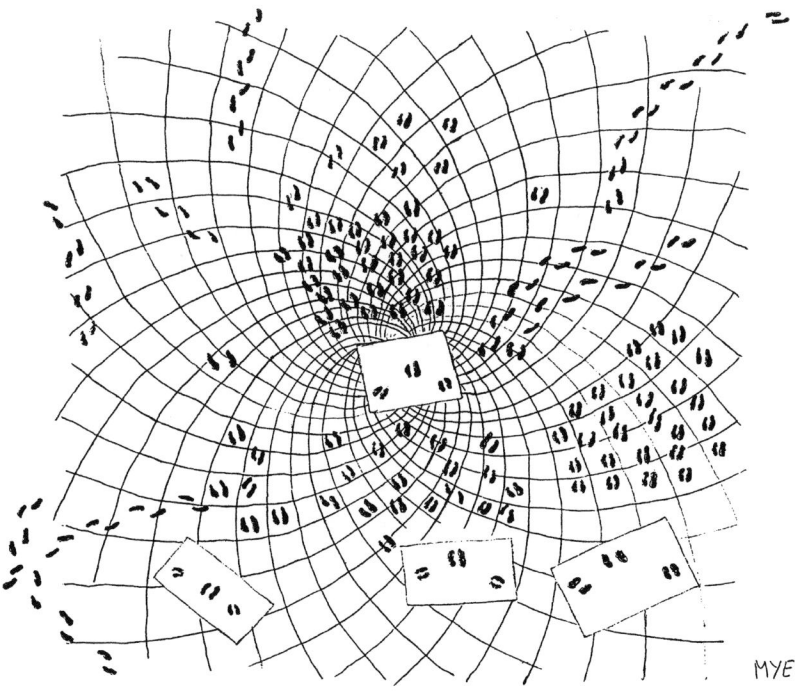

# Évariste and Héloïse

Marco Abate

A few months ago I was asked to propose a story about Évariste Galois, the French mathematician, for the French comic books market. I decided from the very beginning that I didn't want it to be a biographical sketch; I wanted to use the (admittedly interesting) historical character to do something else, something closer to my own personal obsessions. So I started playing around with a few tales I heard, a few people I met, a few books I read … And Héloïse was born, asking for her story to be told. And Auguste Dupin, the original Poe detective, was living more or less at the same time as Galois, a coincidence too good to let pass. And I was also sort of thinking (well, fiddling around might be a better expression for what I was actually doing) about possible relationships between magic and mathematics, and their different but sometimes uncannily similar ways of exploring unknown worlds …

Anyway, given my writing habits, the final product will probably be completely different from what is contained in these few pages; any suggestions for directions I might pursue (or ought not to pursue) will be appreciated, carefully considered, and then thrown away together with the rest of the story when the editor discovers that I'm writing something completely different from what he had in mind when he asked me in the first place. Oh well …

Enough chatter. Let us start by talking a bit about Galois, and the reasons I'm interested in him.

> **GALOIS, ÉVARISTE** (1811–1832) French mathematician famous for his contributions to higher algebra, gave his name to the Galois theory of groups. He was born Oct. 25, 1811, at Bourg-la-Reine, where his father was mayor, and entered the Lycée Louis-le-Grand in Paris in 1823. Despite his genius for mathematics he was evidently a difficult student. He failed twice in the entrance examinations to École Polytechnique; he was accepted at the École Normale in 1830 but expelled the same year for a newspaper letter about the actions of the director during the July Revolution. In 1831 he was arrested for a threatening speech against King Louis Philippe, but acquitted; then shortly afterward he was sentenced to six months in jail for illegally wearing a uniform and car-

rying weapons. He died in May 31, 1832, when only 20 years old, from wounds received in a duel, possibly with an *agent provocateur* of the police.

Galois published only a few small papers; three larger works on the theory of equations were refused by the French Academy. Knowledge of his mathematical achievements stems mainly from a letter to his friend Auguste Chevalier written on the eve of his fatal duel and from posthumous manuscripts.

(*Encyclopædia Britannica, vol. 9, p. 989, 1962 edition*)

(More on this fateful letter later.) Even if he probably won't admit it, Galois is a teenager through and through. A genial teenager, no doubt, but he has all the stubbornness and passion typical of teenagers, only more so. If he gets into something he goes all the way, completely. No stopping halfway, no shades of gray for him. He has no doubts about himself, about being right in whatever he is doing, mathematics or politics or whatever else strikes his fancy (ok, not much else strikes his fancy, but you know what I mean.) He is very passionate, very vocal about his opinions and beliefs, and he lives in a period where being passionate and vocal is both dangerous and exciting. He has the feeling that he is finally solving a problem that eluded the best mathematicians for centuries, and at the same time he believes that his political actions can make a difference to the world – the dream of any teenager. And here lies the first conflict: the world is not paying any (well, hardly any) attention to him. His papers are not recognized: they are lost, or, even if published, fail to get him the attention he craves. He firmly believes in the necessity (and the coming) of a new French Revolution in the name of the Republic; but his efforts produce no results, he is only one of the many young people in one of the many more-or-less-secret societies sprouting up from nowhere in those years. And he doesn't even manage to get recognized as an important member of any of these societies. All his genius, his passion, his efforts – for nothing. Doing mathematics is not a relief for him, it is a necessity. It is both a goal on its own and a means to be recognized by the world. The conflict lies in the distance between the energies he spends and the little he achieves.

There is another aspect of his character that I find interesting: he apparently has no sex life. And this is not standard; for a young man of his age and condition going with prostitutes was absolutely normal. Apparently, he transmutes all his sexual energy (which he certainly has a lot of) into mathematics and politics. In particular, he has a Dionysian approach to mathematics; he feels his work as a physical, sensual struggle against logical difficulties that are very concrete to him. He is no Apollonian thinker tinkering with abstract nonsense; the mathematical world he inhabits is as concrete, as corporeal, as alive as the so-called real world. There is a parallel with his political choices, too. The Saint-Simonian secret society (advocating the government by scientists in the name of reason for the welfare of the people) that he initially associates with is too moderate for him; after a short period he moves to more extremist republican secret societies. He is more interested in doing something for the republic *right now*, without reflecting too much on what happens next. I do not want to stretch too much the analogies between his mathematical attitudes and his political attitudes, but one thing is clear: he has a fully passional, emotional, sensual approach to all important aspects of his life – but sex.

Auguste Dupin, on the other hand, is the embodiment of the Apollonian thinker: no emotion, only pure logic, pure analysis. Which does not mean he does not enjoy using his abilities; on the contrary, he enjoys them exactly because he is not emotionally affected by them (or, better, he affects not being affected.)

> As the strong man exults in his physical ability, delighting in such exercises as call his muscles into action, so glories the analyst in that moral activity which disentangles. He derives pleasure from even the most trivial occupations bringing his talents into play. He is fond of enigmas, of conundrunums, of hieroglyphics; exhibiting in his solutions of each a degree of acumen which appears to the ordinary apprehension preternatural. (…) At such times I could not help remarking and admiring (although from his rich ideality I had been prepared to expect it) a peculiar analytic ability in Dupin. He seemed, too, to take an eager delight in its exercise – if not exactly in its display – and did not hesitate to confess the pleasure thus derived. (…) "But it is by these deviations from the plane of the ordinary, that reason feels its way, if at all, in its search for the true. In investigations such as we are now pursuing, it should not be so much asked 'what has occurred,' as 'what has occurred that has never occurred before.' (…) I said 'legitimate deductions;' but my meaning is not fully expressed. I designed to imply that the deductions are the sole proper ones, and that the suspicion arises inevitably from them as the single result. (…) Now, brought to this conclusion in so unequivocal a manner as we are, it is not our part, as reasoners, to reject it on account of apparent impossibilities. It is only left for us to prove that these apparent 'impossibilities' are, in reality, not such."
>
> (Edgar Allan Poe, *The Murders in the Rue Morgue,* 1841)

(More than a shade of Sherlock Holmes here.) The world has to be stripped down to its basic elements, and reconstructed starting from there. He is rationality's voice (not to be confused with the voice of reason.) He is unaffected by emotion, Galois is pure emotion. Galois is clearly unbalanced; Dupin never falters. Galois has a fully sensual approach to an ethereal subject such as mathematics; Dupin has a purely logic approach to an earthly subject such as murders. A few years older than Galois, Dupin meets him for the first time at a Saint-Simonian meeting (Dupin would have been a Saint-Simonian in his youth, no doubt.) Dupin appreciates the vitality of Galois' intelligence, Galois needs the moderation of Dupin's ways; they become friends.

And then, there are the dreams. Or visions. Most mathematicians, when working, live in a world of their own; Galois (or, at least, my fictional Galois) goes a step further. He actually sees his mathematical world; he enters it. In the dreams/visions he has when working (and not only then) he is able to interact physically with this ideal universe, this universe of ideas. It is (or, at least, he perceives it as) a physical universe, with mass and substance and winds and storms. It has lands and landscapes, boulders and crevasses, infinite stretches of ugly trees ending just beyond the horizon and isolated phosphorescent flowers imploring wanderers to care for them forever. It is shaped by his research, in ways he cannot predict. It directs his own discoveries, in ways he couldn't have imagined.

ÉVARISTE GALOIS ET AUGUSTE DUPIN

**Fig. 1.** Galois and Dupin. © 2005 Paolo Bisi

What I have in mind is a sort of crossing between the platonic world of ideas we mathematicians might be exploring, and the "ideaspace" some philosophers (and some magicians) wonder about, a dimension inhabited by all the concepts humans (and aliens) could imagine, where all stories are true. It exists outside our limited notions of time and space, as fully given as Einstein relativistic four-dimensional space is. A portion of it is devoted to mathematical ideas, and the strength of the sensual approach of Galois to mathematics can take him there.

An important point here is that, even if he has visions, Galois is not crazy. He behaves in all respects as a sane person, a passionate, instinctive, unrestrained basically sane person. He just happens to see something that others don't, exactly as others don't see (or understand) his mathematics. He will endure a lot of stress during the story, and at the end the boundaries between what is real and what is not will be a lot more blurred than at the beginning, but this is in the nature of the story and has nothing to do with his sanity. Galois and Héloïse are not, and won't become, crazy, visions notwithstanding.

In the story I do not want to dwell too much on the philosophical aspects behind this idea of "ideaspace", even though I'm sure Dupin will have something to say about it. I need it and it suits me for four different reasons. First, it gives me a way of visualizing in a metaphorical way both the mathematics Galois is doing and the way he is doing it. I imagine a sort of Fomenko landscape, impossible and horrifying and yet deeply beautiful. It looks hellish to us because we are not equipped to in-

**Fig. 2.** Galois in the ideaspace © 2005 Paolo Bisi (see the section in colour)

151

terpret it, as Euclidean beings immersed in a highly non-Euclidean space. It causes sensory overloads; and yet, we glimpse its inner beauty, its hidden symmetries. It might be a sort of 3-dimensional projection of a much higher-dimensional algebraic manifold (like the way aperiodic tilings are obtained as projection of a more regular 5-dimensional grid), a manifold whose symmetries are given by the Galois group of some equation.

**Fig. 3.** Paris during the cholera epidemic © 2005 Paolo Bisi

Second, the horrifying aspects of Galois' dreams fit very well with the dreadful backdrop of Paris during the cholera epidemic of 1832 (almost 20,000 deaths in less than five months.) They can be deeply related, both visually and metaphorically. In the last part of the story, Galois will go from visions to reality and back almost seamlessly. From the incomprehensible horrors of bodies left dead in the gutters of everyday Paris to the horrors of outlandish landscapes, incomprehensible and yet carrying the promise of being tamed, one day. I don't yet know precisely how we'll get there, but I do know that while lying in the grass waiting to die after the fatal duel, Galois will be in his visions, where he at last … well, I'll tell you later what at last.

Third, the surface detective story will also deal with the ideaspace (I need a surface story to keep casual readers interested. The reason I'd like to tell this story is because of the characters involved; the reason most readers will – I hope – buy it is because of its surface.) One of the elements leading to the fatal duel will be a sub-sect of Saint-Simonians meddling with magic and architecture, trying to get a hold on the ideaspace. In a completely scientific way, of course. It is not as wacky as it sounds; much wackier ideas have been "scientifically" pursued in the eighteenth century (and they still are even in this century, for that matter.)

Fourth, it will give me a way to let Galois and Héloïse interact. Because Galois is not alone in his visions. There is a woman. He doesn't know who she is (she will eventually know who he is, but more about this later.) She shouldn't be there. She is distracting. She is enticing. She cannot be explained away. She is affecting his visions, making them … better? She is frightened, at least at the beginning. But then, slowly, they start to interact, to communicate. And the ideaspace changes with them. It is like a dance, a courting of ideas and souls, modifying their mindscape. He wants her to be there, he longs for her. He will take crucial decisions in crucial moments because of her; and she will learn a great deal from him. And in the end … well, I'll tell you later what will happen in the end.

I'll come back to Héloïse soon. But first I have to give a very rough synopsis of Galois' half of the story. The starting point is the (historical) suicide of Nicholas, Galois' father, in his Paris apartment. Apparently, Galois' father killed himself because of a libel insulting his friends and relatives. The libel appeared under his name, while in reality it was written by a Jesuit who, for political reasons, wanted him to resign as mayor (I'm making a long story short.) Galois, who loved his father dearly, is very much affected by his death; he cannot understand how could he kill himself. And indeed Dupin's position is that he didn't; there is no logical reason for it, the libel is too weak an excuse. The only logical possibility is that he was killed, and they (Dupin and Galois) have the responsibility to discover why, and by whom.

So the idea is to structure Galois' half of the story as an investigation, which also somehow parallels his mathematical investigations. Galois and Dupin will discover that the building where Galois' father lived in Paris was built according to particular mystical proportions and symmetries (which neatly fit with Galois' discoveries about the symmetry group of an equation, and thus with the visual shape of the ideaspace he is traveling in) by a sub-sect of Saint-Simonians, who wanted to use the building as a starting point for the investigation of the ideaspace. To get into the ideaspace, they need to be in a very particular frame of mind; and to reach it, they are ready to use any possible means, starting from the shape of the building up to morbid rituals involving recently deceased Parisian citizens. Of course, they are doing all of this for the eventual benefit of humanity – and because they want to know what is there. (An important and difficult point here is that I don't want them to be "bad guys"; they're doing what they're doing because they think it is the right thing to do. There are often very good reasons for doing opposite, and incompatible, things, and every choice is wrong – or right, depending on which side of the mirror you currently stand.)

Dupin's and Galois' conjecture is that Galois' father was killed because he got in the way of the Saint-Simonians' goals. So Galois wants to try and actually get in their way too, and this will lead to his imprisonment and eventually to the duel.

We follow Galois' investigations, prompted and guided by Dupin. The visions intensify, as well as Galois' political involvement. All his actions become more intense, more passionate – and his frustration grows too. In a meeting of ardent republicans, he proposes his by now (in)famous toast "To the king!", glass in one hand, dagger in the other. The next day he is arrested for defaming the king, his name suggested to the authorities by the Saint-Simonian sub-sect, who didn't appreciate his meddling.

In prison. A living nightmare. The horrifying beauty of his visions his only solace. She is there, too, in his visions. Dupin keeps investigating. Cholera rages throughout Paris. Unexpectedly, this helps Galois: he is released because of city health reasons. He comes out devastated. He keeps going because of his mathematics – and because of her. But she lives in his visions only. He confronts the Saint-Simonian sect, and their reaction leads to the duel. His visions merge with the hellish Parisian alleys ravaged by cholera.

The night before the duel, he writes a few letters.

> All night long he had spent the fleeting hours feverishly dashing off his scientific last will and testament, writing against time to glean a few of the great things in his teeming mind before the death which he saw could overtake him. Time after time he broke off to scribble in the margin "I have not time; I have not time," and passed on to the next frantically scrawled outline. What he wrote in those last desperate hours before the dawn will keep generations of mathematicians busy for hundreds of years. He had found, once and for all, the true solution of a riddle which had tormented mathematicians for centuries: under what conditions can an equation be solved?

(E.T. Bell, *Men of Mathematics*, 1937)

This famous description is historically inaccurate. Galois actually wrote just a résumé of his previous work, with some new remarks and further explanations (and indications of missing details that he had "not the time" to fill in), but the main mathematical core of his researches was contained in the memoirs he had already submitted to the French academy (where they were either rejected or lost.) He also wrote a couple of short letters to friends, briefly explaining the reasons for the duel (which, in reality, had nothing to do with politics or the Saint-Simonians, but was apparently due to some quarrel over a girl: his first approach to sex led him to death. How poignant.)

But here we are not interested (that much) in the historical Galois. We can however use this feverish sleepless night to bring him to the final merging of his own two realities. In an abandoned field at the outskirts of Paris, alone; and in moving waves of unfathomable symmetries, with her. Dupin arrives too late to save him, to tell him how he found out beyond any shade of doubts that Nicholas Galois killed himself, possibly because the peculiar structure of the building he lived in had a magnifying

**Fig. 4.** Héloïse © 2005 Paolo Bisi

effect on his depression. It was no direct fault of the Saint-Simonian sub-sect. The suffering, the prison, the duel: it has all been for nothing. (Or has it?)

Galois' story will be intertwined with Héloïse's story, in such a way that main events in one story will be reflected (distorted, magnified, symmetrized – if such a word actually exists) in the other. Héloïse is about thirty years old, a high-school mathematics teacher in modern day Paris.

155

Yes, I know Héloïse well, nobody better than me, I think. We've been best friends since first grade … A funny little mouse she was – and then look at the beauty she became! But she always kept her mousy ways, somehow … mostly just doing what was expected of her. Even though stubborn to the bone sometimes she was, mind you, when she really wanted something there was no keeping her back. Look at the way she got François. All us girls drooling after him, and who got the prize? Mousy Lise! Unassuming, unpretending Mousy Lise … but she had more than good looks on her side. And I'm not only talking about her brains, even though she had a lot of it, and François had always admired that. She was, you know, a bit wild in bed. Just an eany-teany bit wild, François wouldn't have liked real wild, just enough to get him hooked. And hooked he was! He's always been faithful to her since then, and he could have had all the girls he wanted. Five years older than us, good looks, very good prospects before and that incredible job of his after …

Maybe it was that freak streak in her. Everybody's best girl, respectful, studious, polite, never a worry for her parents or teachers … and then bang! She went and did something completely outrageous, like that, out of the blue. And she was awfully proud of that, whatever *that* was, ready to defend it against the whole world. You know that when she was sixteen she got a piercing on her left nipple? Incredible. And her mathematics! She wanted to be a mathematician, for heavens' sake! A *mathematician!* Can you believe that? Not even François could dissuade her – well, eventually he did, but you know what I mean, their daughter wasn't exactly what you'd call expected … She had to cope with that, and she did, you have to grant her that. François was wonderful; they got married right away, and then he bought that wonderful apartment … he is a big shot now, you know, in that financial firm, what's-its-name … a lot of work, but he *is* good, I can tell you. Why Héloïse wanted to go teaching is over my head. With her house, and her husband, I'd just stay home the whole day … and night … It must have been that freak streak of her. But she is a very good mother, I grant her that, and they tell me she is a good teacher too.

No, I don't see her that much anymore. With her work, and her daughter, and her house, and her husband, she has no time for her old friends, or at least so she says … Happy? Is she happy? What kind of question is that? Of course, she is! *I* would like to have her house, and her husband! Look at the place I am forced to live now; this is not what I expected when I met Henri… Héloïse got François first, and Henri was his best friend then… She got the better deal, she did, don't get me started about Henri, you'd better not… She'd better be happy! Don't you think so?

(Excerpts of an interview with Denise Marchand, Ici Paris, 2004)

Héloïse lives, together with her husband and her six-year-old daughter, in the same apartment Galois' father lived in. She wanted to become a professional mathematician, and she was quite good at it, but during grad school she got pregnant, and you know how these things go. She found herself at home caring for the baby while her soon-to-be husband worked his way up in a financial firm. As soon as she could, she got a temporary position as a high-school teacher, but it is not what she had hoped to become. Her life is shades of gray. She is a good teacher, she is a good mother, she tries to be a good wife – and this leaves no time for her own needs and desires. Which is something her husband doesn't understand, doesn't even imagine could be a problem. In his own terms, he thinks he is a good husband; he just doesn't get it.

Héloïse (possibly because of the building they live in) has a recurrent dream; not often, but often enough. She dreams of another, incoherent, formless, horrible yet beautiful place. With possibly somebody else there. A shadow. An incoherent formless alluring shadow.

This is her life: her daughter, her work, her house, her husband, sometimes the dreams. And then an old friend, Antoine, shows up.

**CHEVALIER, ANTOINE** (1973) Born in Toulouse on May 4, 1973, of Louis Chevalier, journalist, and Marie Lagrange, secretary in a law firm. A brilliant student, he was accepted at the École Normale in 1991. There, he received his *doctorat* in Mathematics in 1998, under the direction of Prof. François Berteloot. His ground-breaking work on the application of novel algebraic techniques to the study of holomorphic dynamical systems won him a Post-Doctoral fellowship at the University of California, Berkeley, for the years 1999–2000, and then a 4-years position as assistant professor at the Princeton University. In 2005 he accepted a permanent professorship at the École Normale Supérieure in Paris; he is now one of the youngest professors at the Institut. Besides doing mathematics, he likes walking the streets of Paris at night, and folding unreal origami shapes.

(*American Mathematical Monthly*, vol. 127, 2005, p. 327)

Antoine and Héloïse were grad students together, then she got pregnant, his postdoc took him to the States, and they lost contact. After a few years abroad, he landed the position in Paris, and now here he is. They meet, they start talking, and it is like the most natural thing, as though time hadn't passed. He inquires about her mathematics (which had something to do with Galois theory); she was doing quite a good work at the time, and since then nobody else has gone in the same direction which now looks even more promising than before, why doesn't she give it another try? She must complete her work; she can do it!

She hesitates, but eventually she decides to try. And as soon as she reenters mathematics, she realizes how much she missed it. To be in a world she feels as her own, where success can be achieved, if ever, by her own efforts only, where she doesn't depend on anybody and nobody depends on her ... But doing mathematics requires time, time that must be taken off the rest of her life. Her husband doesn't understand, he feels rejected, he feels jealous – their sex life suffers.

At the same time, the dreams become stronger, more structured. She is afraid, and attracted. She is no fool; she soon understands the connection between coming back to mathematics and the dreams. The formless presence she sensed before becomes more and more concrete, and finally she recognizes him, Galois. They do not talk, in the dreams; they communicate by manipulating the landscape in ways they later translate into mathematics (and ideas, and emotions). She knows his story (all mathematicians do), and she knows he is going to die soon. Or is he already dead? After all, it is only a dream, isn't it? It just doesn't feel like a dream anymore.

157

> One way to look at rationality in dreams is to classify different levels of lucidity. At the highest level, the dreamer would not only be aware of dreaming, but also possess complete understanding of the implications of this knowledge, and would behave in accordance with that understanding on all levels from thought to action. The lowest, minimal level of lucidity would be realization of dreaming, but without understanding how dreaming is different from waking, and without acting on the lucidity at all, mistaking events, characters and consequences with those from waking life. Yet, degrees of rationality vary from moment to moment in dreams, so that one wishing to use a scale of levels of lucidity would have to rate each decision, action, or response of the dreamer independently. Averaging the lucidity levels in a dream might be a way of establishing a lucidity "score" for the dream. All of this is for future research to decide.
>
> (Lynne Levitan, *A Fool's Guide to Lucid Dreaming*, NIGHTLIGHT 6(2), Summer, 1994)

Héloïse accepts the dreams, as she has accepted all that has happened to her in her life: her daughter, her husband's decisions, Antoine's suggestion of going back to mathematics ... It's one of the reasons I think Galois and Héloïse complement one another so well. Galois is all action, he tries as much as he can to steer his life in the directions he want. Héloïse, instead, is led by her life; she tries to make the best of what happens, but she almost never tries to make things go in the directions she'd like them to go. On the other hand, Galois is unable to organize his life and his efforts in a productive way, while Héloïse knows all about organizing things (as almost any woman who works and is also a mother and a wife does.)

She cannot talk with her husband about the mathematics, or about the dreams either. They become more and more estranged, even though on the surface apparently nothing has changed. And then it happens. The husband is away, Antoine is there, they start talking, drinking, they become intimate, they end up in bed.

She finds in him something that up to that moment she almost didn't know was missing. Her mathematics, her lover … she now has a life of her own, it doesn't matter what is going to happen, it doesn't matter how difficult it is, this is what she needs now. Even though managing her life has become a nightmare. Keeping all the elements of her life together and separated at the same time is almost impossible, but she doesn't know how (and doesn't want) to leave anything behind.

Then, possibly because of the properties of the building, her husband discovers something – and a crisis is precipitated. He doesn't want to lose her (or to give her more space either – she's always had, *I*'ve always given her enough space, haven't I? I didn't do anything to deserve this!) He just wants things as they were. He throws away her mathematics. He doesn't know what he did wrong – he honestly tries to understand, but he cannot. Again, he is not a bad person, he is trying to do his best; but sometimes the best leads in the wrong direction. And I don't want to be preachy; I hope to portray them as realistically as I can, with their strengths and their weaknesses, and not as role models to follow or despise.

In a particularly harsh row, he hits her – for the first and only time. She falls down, loses consciousness, and finds herself in the dream. Galois is there; he is dying, hit by a bullet in the chest, the morning of the duel. She approaches him; they touch. They embrace, they make love. Giving, at last, to somebody who understands. Content, in a moment they both want to be in. She gives him recognition at last, in the fully corporeal instinctive and fulfilling way that only a woman can give to a man; and he gives her the strength to steer herself at last (and no, I'm not going to say "as only a man can give to a woman", because first it is not true, and second, one should not stretch one's rhetorical arguments too far; but Galois *is* the right person to give *her* that strength in that precise moment.) At the end of his life, Galois finally obtains what he most desired, and he gives her what she most needed then. Maybe it is a vision, or just a dream, or maybe it is actually real, in an outlandish Fomenko-Little Nemo way, it doesn't matter; his life ended as he wanted it to, and her life started again, as she hadn't let herself even imagine it could.

She awakens in a hospital: head concussion, nothing serious. When she leaves the hospital, she leaves her husband too, taking her daughter with her. She doesn't move in with Antoine, at least not yet. She wants to be in charge of her life, now; she's dreamed of it long enough. And she leaves something else in the hospital; she consciously decides to. She doesn't need it anymore. Her piercing. It was on her left nipple. Over her heart. In the shape of a small, but perfectly recognizable, bullet.

# Mathematics and Animated Cartoons*

Gian Marco Todesco

Animated films are an extremely flexible art form, as they allow the filmmaker complete freedom of expression. There are no restrictions relating to the laws of perspective or physics, nor to the form or appearance of the characters. Rarely will the audience, immersed in a totally fantastic world, realise how much work and exact organisation are necessary to give fantasy shape. However, as we shall see in the following, the realisation of an animated feature film is a gigantic piece of work that demands an immense commitment. It is not surprising that during the last fifteen years the computer started playing an ever larger role in this field. Obviously, this has required a number of dedicated programs to be invented and developed. All over the planet software houses have specialised in this sector, however, the programs most widely used at the professional level are less than five. One of these programs, Toonz, is produced by an Italian company. The author of this chapter is one of the creators of the program, which allows him to observe the world of animation from an unusual perspective. The following talk aims at illustrating this perspective and with a few examples it shall reveal the mathematics and computer science that work, unseen, behind the scenes.

## Toonz and Traditional Animation

In its almost twelve years of existence Toonz (see Fig. 1) has been used by hundreds of companies all over the world in the production of short films, commercials, interactive games, television series and feature films. Here I will cite but three of the best known full-length films.

> *Balto* (Simon Wells, Amblimation, 1995) was the first film to be produced with our software. The program was still in the test stage and during the production

---

* *Translated by Sarah Wolf*

there was an intense and fruitful interaction between our group and the experts from Amblimation. The stage at which the artist cannot yet imagine the possibilities the program has to offer, while the computer expert does not have the slightest idea of what the artist considers useful is splendid. Finding a common language in which to reach an understanding was at the same time difficult and highly satisfying.

*Anastasia* (Don Bluth & Gary Goldman, Fox Animation, 1997) was another milestone. Shot in cinemascope format, the film has led to the management of "enormous" images in terms of the required memory and calculating capacity. Often the background sizes were up to 100 times larger than a common digital photograph.

Finally, I can proudly mention the film *Spirited away* (Hayao Miyazaki, Studio Ghibli, 2002), winner of the Golden Bear Award in Berlin 2002 and of the Oscar as best animated film in 2003. In 2005 the director was awarded a Golden Lion for lifetime achievement at the Venice Film Festival. His film *Howl's moving castle*, released in Italy in the same year, was also produced with Toonz.

All examples mentioned were realised using the traditional animation technique, *cel animation*. The term derives from the transparent sheets of celluloid or acetate onto which the drawings used to be photocopied before the introduction of digital technology. For these films the artist draws the characters by hand, often using traditional tools such as pen or pencil. Today the term animated film is used in a much wider sense, including 3D animation. In a 3D film the characters, objects and backgrounds are not actually drawn, but are instead created virtually as three-dimensional models which the computer is then able to "photograph" and "film" applying the laws

160

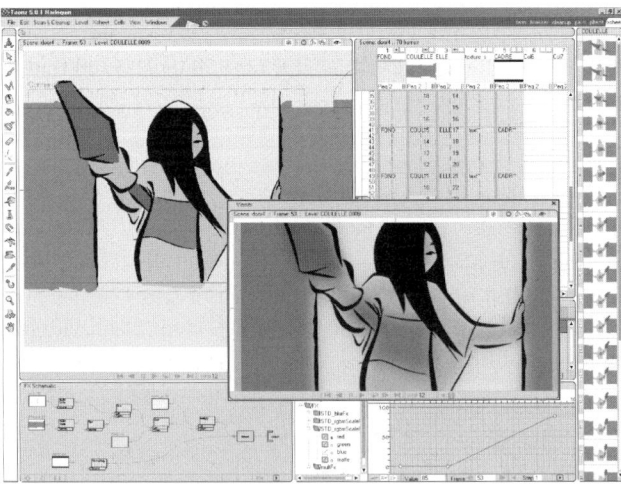

**Fig. 1.** The program Toonz 5.0 Harlequin, by kind permission of Digital Video s.p.a., http://www.toonz.com

of perspective and optics. This technique, which has made possible the creation of wonderful movies such as *Toy Story* (Pixar, 1995) or *Shrek* (Dreamworks, 2001) is completely different from the former and, fascinating though it may be, will not be considered here.

## The Production Process

The production of an animated feature film involves hundreds of people and is rigidly organised in different stages. The general scheme has emerged steadily right from the beginning of animation, when production was completely manual. The introduction of the computer did not affect the proceeding deeply, it merely modified the manner of realisation of some stages. In the following paragraphs I will give a short (and very much partial) summary of the most important steps of the process. The real procedure is much more complex and varies greatly from studio to studio.

First of all the story board, a sort of drawn film-script of the subject matter, is created. The story board gives an account of all scenes of the film, as in a very long comic-strip, containing framings, camera movements, and information on music and dialogues.

Next, the appearance of the different characters is defined. Everyone is summarised in a model sheet that portays him with different poses and expressions. Using as basis these model sheets, which guarantee stylistic consistency of a character over the span of the whole film, the animators draw the long sequences of images which are the essence of animated films. For each movement of the character, the animator creates a few fundamental frames: the beginning, the end and possibly other salient movements. An assistant uses these pictures as a guide and adds intermediate images. In their turn the *inbetweeners* add other images in order to arrive at the necessary number of frames per second: usually 12 or 24.

The soudtrack has been recorded in advance and synchronisation between the voice and the movement of the characters' lips is assured by a sort of reverse dubbing: the drawings follow the talk and not vice versa. Then, if a foreign film arrives in Italy, it undergoes a true dubbing which obviously worsens the syncronisation.

Further, the drawings are inked. This stage is one of the most labour-intensive ones, and was particularly such before the use of the computer. Colouring is a task which demands great patience and precision, but which can be delegated to workers who are not particularly qualified. It was common practice to send the reams of sheets to the other end of the planet, to countries with a considerably lower labour cost, with drawings to be inked and precise colouring instructions to be followed. Today with the computer one click suffices to precisely paint in a closed area giving it the selected colour. It is also possible to

161

automatically colour a certain number of correlated areas. Nevertheless, this activity still requires many working hours of a human worker. Today the sheets do not travel any longer, but still in their stead the bits representing them are transmitted.

The backgrounds are produced by real painters, using a great variety of techniques: watercolour, acrylic or oil paint, crayons, collage, etc.

Finally, everything has to be put together. Each frame of the film is made up of several layers or different levels: the background, the characters, the shadows, possibly atmospheric effects, etc. Without the computer it was necessary to photocopy each picture on a transparent sheet, place one sheet on top of the other and photograph them with a special animation camera, frame by frame.

The information which allows the cameraman to know which components are present in each frame is collected in a scheme called exposure sheet. The exposure sheet also contains the instructions which control the movement of the camera, the framing and the relative positions of the various levels. It is common practice to have a background which is much larger than the framing and to make it slide in the course of animation so as to simulate a dolly shot.

To sum up the process quantitatively: a film of two hours, with 24 frames per second, consists of more than 170,000 frames, each of which, as we have seen, results from the composition of different elements. All in all up to 300,000 drawings may be necessary.

162

## Mathematics and Computing Science Enter the Scene

Let us now see in which production stages the computer can intervene. Different from what one might think the stage of drawing characters, frame after frame, is still realised by hand and on paper. Only in the last years have there been faint signals of change, related to the development of a new generation of graphic tablets, to the diffusion of cartoons on the web and to the generational change of animators who nowadays have a greater familiarity with the computer. Nevertheless, today, in the production of an animated feature film, the drawings are still done one by one in pencil. These drawings are then scanned and transformed into digital images. From this point onwards the whole production process is digital, up to the printing of the film.

The software thus has to control the scanner, put the finishing touch to the pictures (perhaps compensating automatically for stroke differences between one drawer and another), colour the drawings, define and modify the exposure sheet, add lighting effects, transparencies, defocus transitions, etc. and eventually assemble all the components to generate the final frame. A whole series of related problems go along with these fundamental tasks. It is essential to have instruments at hand to find one's way through the vast database of scenes, characters and colours; it has to be possible to rapidly generate test films to control the smoothness of the animation even before the drawings are inked; one has to manage the so called *render farms*, a group of powerful computers linked in a network, to which the burden of work related to the creation of the final sequence is distributed automatically; etc.

A software which provides all the necessary instruments ends up being an extremely complex and articulate object. The source code, that is, the set of instructions that form the program, can amount to several hundreds of thousands of lines.

The programs for producing animated films, just as every other program, presuppose a series of models which act as a bridge between the dull precision of the computer and the real world, ever changing and hard to define. These models generally make ample use of the language of mathematics. It is impossible, here, to provide a complete or even only an indicative list of the various mathematical models that are employed, but I hope that the four examples presented in the following paragraphs are sufficient to convey the basic idea. We will examine the form of the brush strokes that compose a single drawing, the management of the camera movements in composing the final scene and lastly the digital generation of some special effects: clouds and the so called particle systems.

## The Virtual Paintbrush

The great majority of animated films of a certain quality pass through drawings on paper, but things are slowly changing. The artists are starting to have a less hostile attitude towards the computer and a new family of graphic tablets with built-in screen allows a more natural way of drawing. Presumably, the percentage of entirely digital animated films, at present outnumbered, is bound to grow in the years to come.

Obviously, the software has to adapt. If the drawing comes about directly in the computer, it is convenient to create a representation of it which is more articulate and flexible than simply to retain the pixels. Hence, to this end we need a geometric model which represents the single stroke of the brush. Keeping a geometric representation of the brush stroke permits improvement of the drawing process, allowing for modifications and corrections that would otherwise be impossible. Moreover, as the image produced is not based on pixels, it is independent of the resolution (in other words of the number of pixels available) and can be used both in the field of television and in that of cinema without any loss in quality.

When designing the form of the brush stroke we have to keep in mind that the pen of a grafic tablet, a bit like a real brush, is sensitive to pressure and allows to continuously vary the width of the mark. Thus it is possible to create a particularly expressive and personal stroke. Apart from this characteristic, our "geometric" brush stroke has to have smooth contours without visible junctions, it has to be described by computationally light formulas (to allow the computer to draw in real time the thousands of strokes that form a drawing) and it must be perfectly defined by means of a reduced set of control points (see Fig. 2).

**Fig. 2.** A picture with some control points displayed

The mathematical model that serves our purpose is the so called *spline*, a very useful and versatile parametric curve, invented on purpose for CAD (computer-aided design) programs. It is a continuous and differentiable curve which roughly follows the pattern of a given piecewise linear curve, or traverse: to completely define the curve it suffices to specify the vertices of the traverse, the so called control points. In our case the curve has three dimensions: the first two are relative to the drawing plane, while the third one represents the width. A series of algorithms automatically generate the control points, without the drawer having to worry about them. Also, at the stage of corrections and modifications the control points are created, moved and deleted in an automatic manner, while the artist simply moves, drags or twists the curve as if it were a string or a piece of wire.

An interesting characteristic of this instrument is the possibility to associate an arbitrarily complex "style" to the curve itself. Thus it becomes possible to simulate a series of conventional tools like pencil, charcoal or watercolour and it is also possible to invent completely new instruments, as shown in Fig. 3.

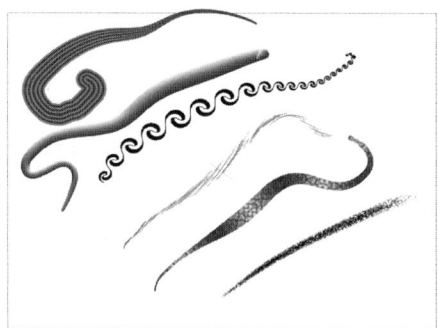

**Fig. 3.** Different "styles" of brushwork

## Transformations in Motion

The elements that compose the scenes of a cartoon can move with respect to each other. For example, the background can slide sideways like the countryside outside the window of a train in motion. A sequence of a few frames representing a bird that flaps its wings can be repeated many times changing the position of the animal with respect to the background to establish a long continuous flight from one side of the screen to the other. A third and particularly important example are the so called camera movements, one of the main tools at the director's disposal for animated films just as in all other types of film. In this case what moves is the framing, that is, the rectangle which specifies what part of the scene will be "shot".

In all these cases there are parameters (e.g. the size of the framing or its position) which can vary as a function of time. The variation can happen at a constant velocity or with accelerated or decelerated motion, depending on the director's instructions. As the following example illustrates, controlling these variations appropriately presents more than one subtlety.

Let us examine a very natural motion: "zooming in" or in other words the progressive enlargement of the image which to the spectator gives the impression of approaching the depicted object. The first scene of the film *Bambi* (Walt Disney, 1942) is a long zooming which starts off from the forest and closes in on the lair of the fawns. More recently, the film *The Hunchback Of Notre Dame* (Walt Disney, 1996) concludes with a very long zooming out which departs from the celebration in the square showing the whole church and then the entire city. Our more modest example also centers on this form of motion.

To make this example interesting we will consider a very curious subject: an engraving by the Dutch artist M.C. Escher: *Smaller and smaller* (1956). This work presents a symmetry, like very many of the artist's works do. In this case it is a symmetry of scale: if I enlarge the image by a certain factor and overlay it with the original, the central parts coincide almost perfectly.

In fact, in the engraving, the basic motif forms a frame which is repeated many times, every time reduced in size and closer to the center. The smallest frames, closest to the middle, present details which are almost indistinguishable to the naked eye, however, inside there would be space for infinitely many ever smaller frames. As a whole the image gives the impression of a long perspective flight towards an infinitely far away central point.

It is this interpretation to suggest the idea of realising a small animation which creates the illusion of a continuous approaching towards the center. The animation is simple: it consists in framing the central part of the artwork (while eliminating the borders which, probably for aesthetic reasons, follow a slightly different pattern) and progressively enlarging the image. The factor of enlargement will be chosen in such a way that the first and the last picture of the sequence are (almost) superposable. We have to determine a formula which allows to find the enlargement factor for each frame. The most natural choice is a linear relation between time and enlargement, that is: $S(t) = At + 1$, where $t$ is the time that has passed from the start of the sequence while $S$ is the function which provides the desired enlargement factor to be applied at time $t$. We have adopted the convention that an enlargement of 1 means to leave the image unchanged and, in fact, we see that $S(0) = 1$, that is, we start with the original image.

The constant $A$ is chosen in such a way that, at the end of the sequence, $S$ assumes the value of the relation between the sizes of two successive frames in the image. At this point we can generate the sequence: this means producing a certain number of pictures, say 50, all obtained by taking the original image, enlarging it by a factor $S(t)$ and choosing the same central rectangle (which thus will show a smaller and smaller region of the original image.)

The so generated sequence is effectively periodic: the first and the last picture are very similar. If the sequence is reproduced in a continuous cycle we have the impression of an uninterrupted movement towards the perspective flight. We realise, however, that something does not work. The movement is not uniform, but slowed down and the repetitions of the cycle generate the impression of a progression with jumps. We have to find out what has gone wrong.

The key to the problem lies in the multiplicative nature of the scale operator. It is intuitive that tripling the size of an image and then doubling the result has the same effect as enlarging it six times $(3 \times 2)$ and not five $(3+2)$. Let us see what this implies in our example. The various frames which constitute the initial picture are each smaller than the preceding one by a certain scale factor $s$. If the outermost frame is of size $L$, the following one will be of size $L/s$, the next smaller one of size $L/s^2$ and so on. At the end of the first cycle of my animation I will have applied an enlargement equal to $s$ to the image, such that each frame occupies the space of the preceding one. In particular, the third frame has taken the space of the second, which in its turn has become as big as the first one was. The outermost frame has become so large that it does not fit into the picture anymore and has vanished from sight. If I suppose to continue for another cycle, I expect the third frame to take the place that the first one had initially. In other words, I want the enlargement to be $s^2$, and certainly not $2s$. Therefore the correct form of the function $S(t)$ has to pass from 1 to $s$ in the same time that it passes from $s$ to $s^2$. The exponential function is an obvious candidate to this end. In fact, the function $S(t) = e^{Bt}$ has the property that the relation between $S(t)$ and $S(t+\Delta t)$ is always the same for a given $\Delta t$. If we regenerate the sequence with the new relation between time and enlargement and reproduce it in a continuous cycle, we observe the desidered uniform progression.

The image considered here is not a drawing in perspective and, in fact, a real perspective flight (for example a drawing of a long corridor, with a series of doors at regular intervals) would present a slightly different disposition of the "frames". Though the formulas that define the two cases are different, one is an approximation of the other. In practice, in the world of animation, having to "zoom in" on a background, one always uses exponentials to control enlargement as a function of time.

# The Shape of Clouds

Apart from proper characters and backgrounds there are many other elements that populate the scene. For example, the so called special effects: rain, snow, lightning, and so on. In many cases, the computer can be used to create these effects automatically and with a certain level of realism. As usual, in order to design the algorithm one must first create an adequate mathematical model.

A representative example is given by clouds or vapour. What is the exact form of a cloud? This question is only seemingly provocative. Of course a cloud is the ever changing and undefinable form par excellence, however, it is also true that given different images of an appropriate colour some will be more "cloud-like" than others. Obviously, the contours must not be well defined and the shape has to be extremely irregular and chaotic. Nevertheless some internal coherence, which a totally random conglomeration of light and dark pixels does not possess, is required.

A technique called *Perlin Noise* enables us to find the right mixture of randomness and order that is characteristic of a vast range of natural phenomena from patterns in marble to a cloudy sky.

Let us begin by dealing with a one-dimensional problem: suppose we want to draw a mountain range. In other words, we want to find a function the graph of which resembles the crest of a mountain. The first ingredient to our recipe is a pseudo-random number generator. This is a particular algorithm able to generate sequences of numbers that have a statistical distribution which is very close to

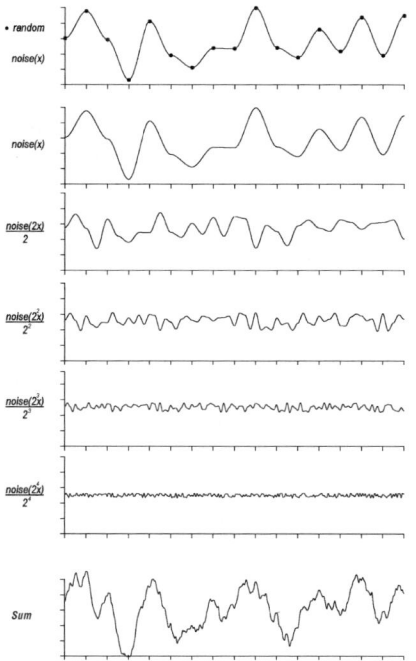

**Fig. 4.** The function Perlin Noise is constructed as a sum of so called octaves – each noise function is twice the frequency of the previous one, like octaves in music. The black dots in the first graph are controlled by a sequence of pseudo-random numbers. The function $noise(x)$ is obtained by interpolation

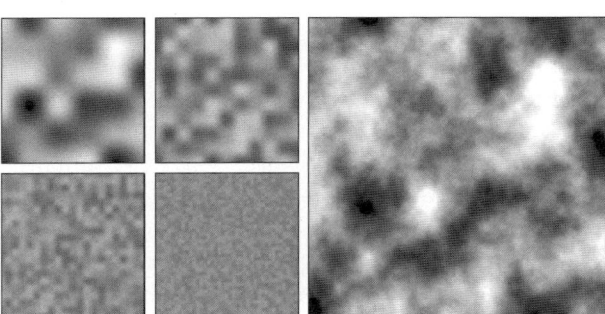

**Fig. 5.** Perlin Noise in two dimensions: the pictures *on the left* show the graphs of four successive octaves. Their sum produces the enlarged picture *on the right*

that of a totally random sequence of numbers, produced for example by rolling the same die many times.

With this instrument and interpolation techniques we can generate a continuous graph without cusps that has an apparently random pattern. The sequence of pseudo-random numbers defines the values of the function corresponding to the respective integer coordinates of the x-axis.

One can generate a family of functions similar to this one by adjusting the amplitude (that is, the maximal range of the values) and the frequency (that is, the speed at which the function changes), in other words, by modifying the scales of the two coordinate axes. Perlin Noise is obtained by summing up many of these functions of increasing frequency and decreasing amplitude (see Fig. 4). This procedure generates the scale invariance characteristic of natural phenomena. In fact, the graph of the function resembles a mountain range with peaks and valleys in irregular alternation; along the crest one can identify smaller peaks that in turn show smaller wrinkles and so on.

The two dimensional analogue of this function makes it possible to create a cloudy sky. Consider the function $f(x, y)$ and assume that its value indicates the brightness of the pixel positioned at the coordinates $(x, y)$. If the function $f$ is constructed as in the previous one dimensional example, the graph of $f$ will show a random pattern which possesses the scale invariance that has just been discussed. With an appropriate choice of colours the image will have the credible appearance of a cloudy sky, as shown in Fig. 5.

## Particle systems

The computer lends itself rather well to carrying out relatively simple tasks that are repeated a large number of times (whereas this type of operation is particularly unpleasant to human beings). An interesting example is encountered in the digital modelling of natural phenomena that one can imagine to be constituted of a large number of basic elements. For example, the flight of a flock of birds or the jet of a fountain made up of countless drops of water. Or further, rain, snow, falling leaves, etc. To represent these phenomena it is neccessary to draw every drop, leaf, bird for each frame. From one frame to the next one has to update the position of each element in the appropriate manner: each bird will flap its wings and move forward, the leaf will slide towards the ground with the characteristic oscillating movement. Given the large numbers of elements involved it is reasonable to suppose that the size of each single element is going to be small and that, in any case, the spectator's eye will rather be prepared to grasp the whole than to concentrate on the single parts composing it. This entails the possibility to use very simple drawings and instead concentrate every effort on the movement.

The part of the program which manages this type of effect is generally called "particle system", where the term "particle" refers to the single element (the bird, the leaf, etc.). To use the program one has to define the image or the images for the various particles and then code the rules which every particle has to follow in its evolution. At every instant the software makes a certain number of particles appear in determined positions and then controls the movement in the following frames. Generally, every particle has a "medium lifetime", past which it will be removed from the system (for example a drop of rain "lives" only until it touches the ground).

During its lifetime, the particle is represented by a picture or a sequence of pictures, possibly cyclic (for example, a bird flapping its wings). Moreover, the program updates the position and the speed of the particle utilising a series of rules established by the user. These rules are chosen in such a way as to simulate the action of forces like gravity or the wind. The velocity, the rate of birth and death of the particles and all other relevant parameters are usually subjected to a perturbation (implemented by means of a random number generator), to avoid that the movement appears too "mechanical" and ordered. Obviously, the importance of this perturbation can be increased or diminished as one pleases.

As we have seen, there can be three different types of particles: relatively large and detailed particles, as for example animals in flocks, herds, swarms and so on, or leafs or petals of a flower which fall. In these cases generally the particle is made up of more than one image, so as to give rise to a small animation: the bird flapping its wings or the leaf turning around. One can also associate more than one animation cycle to a particle, such that the system can choose randomly which one to utilise.

In the second case, the particle is very simple. The idea is to have a large number of them and make them move in such a hurry that it becomes difficult to appreciate any details. Examples are single drops of rain or snowflakes or the luminous spots that constitute fireworks.

Finally, we have particles without a well-defined shape, often with soft contours and of semitransparent design. In the final image these particles cannot be distinguished from each other, but they contribute to creating the image of, for example, a flame, an explosion, a column of smoke, etc.

Particle systems are truly powerful but at the same time complex to utilise. Unlike simpler effects, they demand a minimum of knowledge in physics and mathematics from the user. It is obviously possible to realise a certain number of ready made basic models, which cover most cases: atmospheric effects like snow and rain, explosions, flocks of birds and so on. However, their extreme versatility al-

169

lows a worthwhile usage of particle systems also in scenarios that are completely unforseen by the creators of the program, like the scene of the dragon which crumbles into a myriad of scales in the film *Howl's moving castle* (Ghibli, 2002). This is a good example of a happy synthesis between technical ability and fantasy.

## Web References

GianMarco Todesco, Home Page
*http://www.toonz.com/personal/todesco*

Digital Video s.p.a.
*http://www.toonz.com*

Balto
*http://www.imdb.com/title/tt0112453/*

Anastasia
*http://www.foxhome.com/anastasia/main.html*
*http://www.imdb.com/title/tt0118617/*

Spirited away
*http://www.spiritedaway.com.au/*
*http://www.nausicaa.net/miyazaki/sen/*

Traditional animation
*http://en.wikipedia.org/wiki/Traditional_animation*

Perlin noise
*http://freespace.virgin.net/hugo.elias/models/m_perlin.htm*
*http://mrl.nyu.edu/ perlin/*
*http://astronomy.swin.edu.au/ pbourke/texture/perlin/*

Spline
*http://www.doc.ic.ac.uk/ dfg/AndysSplineTutorial/*
*http://mathworld.wolfram.com/Spline.html*

The sites were visited on June 4th, 2007.

# Mathematics and Art

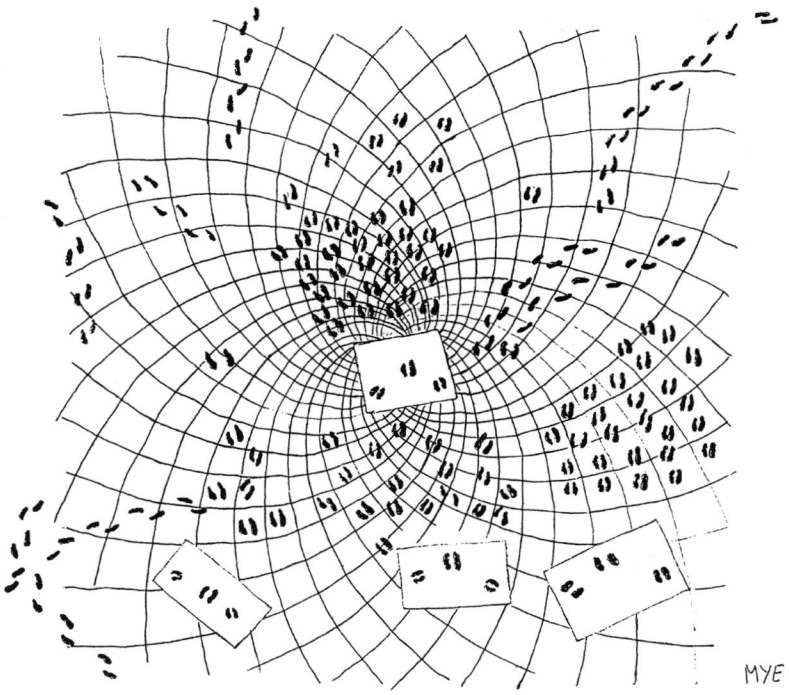

MYE

# Abstraction*

Michele Emmer

## Abstraction: Kandinsky

The mathematician Robert Osserman writes:

> Abstraction functions in various ways. Above all it possesses the power of universality that permits applying a single rule in circumstances that are very different [...] it also has the great advantage of permitting our imaginations a lot of liberty, allowing us to come up with new and alternate versions of reality; versions that may or not correspond to something in the real world [1].

The great message sent by mathematics from the end of the 1800s to the beginning of the 1900s is that geometry, space, and mathematics can be the realm of freedom and imagination, of abstraction and rigour. Geometric objects and mathematical ideas are of universal interest, at everyone's disposal: to non-mathematicians in general, of artists, writers, musicians; to be used, understood, misunderstood, mutated, distorted; their impact essential and, at the same time for the non-mathematicians, esoteric and mysterious. It may seem like a contradiction, but it isn't.

The new epoch begins at the turn of the twentieth century with the publication of the book "Concerning the Spiritual in Art" by Kandinsky, finished in 1910 and published in December 1911, but with the date 1912. "It is a singular privilege for a book that is ahead of its time, to be ahead of itself", comments Elena Pontiggia.

According to Kandinsky, Cézanne had taken still life to a level at which things that were externally dead became internally alive, giving them a chromatic expression, that is, a dimension that was intimately pictorial, and closing them in a form that could be translated into abstract forms, often mathematical, that exude harmony. Harmony is a word that obviously recalls music, and it is to music that Kandinsky turns.

> A painter who finds no satisfaction in mere representation, however artistic, in his longing to express his internal life, cannot but envy the ease with which

* Translated by Kim Williams

music, the least material of the arts today, achieves this end … naturally seeks to apply the methods of music to his own art. And from this results that modern desire for rhythm in painting, for mathematical, abstract construction; for repeated notes of colour, for setting colour in motion and so on [2].

And he adds that the form, even if it is completely abstract and resembles a geometric figure, has an interior sound:

This essential connection between colour and form brings us to the question of the influences of form on colour. Form alone, even though totally abstract and geometrical, has a power of inner suggestion. A triangle (without the accessory consideration of its being acute-or obtuse-angled or equilateral) has a spiritual value of its own. In connection with other forms, this value may be somewhat modified, but remains in quality the same. The case is similar with a circle, a square, or any conceivable geometrical figure. (The angle at which the triangle stands, and whether it is stationary or moving, are of importance to its spiritual value. This fact is specially worthy of the painter's consideration.) [3]

Thus a contrast is delineated between form and colour. A yellow triangle, a blue circle, a green square and, again, a green triangle, a yellow circle, a blue square, etc., are all very different and have very different effects.

It is evident that many colours are hampered and even nullified in effect by many forms. On the whole, keen colours are well suited by sharp forms (e.g., a yellow triangle), and soft, deep colours by round forms (e.g., a blue circle). But it must be remembered that an unsuitable combination of form and colour is not necessarily discordant, but may, with manipulation, show the way to fresh possibilities of harmony.

Since colours and forms are well-nigh innumerable, their combination and their influences are likewise unending. The material is inexhaustible [3].

Just as in the different geometries, it occurs to us to add.

But in spite of the different possibilities, the form cannot avoid two extremes. It is either:

- delimited, which serves to make a material object stand out on the surface, that is, to draw it;
- abstract, that is, it doesn't represent any real object. These purely abstract entities, which as such have life and exercise an effect, are the square, the circle, the triangle, the rhombus, the trapezoid, and the innumerable other, increasingly complex forms that don't have a particular mathematical name. All of these forms have an equal right to citizenship in the realm of the abstract.

Between these two extremes lie the innumerable forms in which both elements exist; with a preponderance either of the abstract or the material.

Kandinsky adds:

> The more abstract is form, the more clear and direct is its appeal. In any composition the material side may be more or less omitted in proportion as the forms used are more or less material, and for them substituted pure abstractions, or largely dematerialized objects. [...] Must we then abandon utterly all material objects and paint solely in abstractions? This is a legitimate question, which can be answered by considering the agreement that exists between the two components of form: objectivity and abstraction [4].

And here the theme of liberty explodes in Kandinsky's words:

> This way lies today between two dangers. On the one hand is the totally arbitrary application of colour to geometrical form – pure patterning. On the other hand is the more naturalistic use of colour in bodily form – pure fantasy. Beyond these limits we find, to the right, pure abstraction, that is, a more radical abstraction that the geometric one, and to the left pure realism, that is, a more vivid fantasy, translated into concrete material. Between these two extremes there is infinite liberty, a depth, a space, a wealth of possibility: everything, today, is at the disposal of the artist. It is a moment of liberty that can only be born at the dawn of a great epoch [5].

And obviously he refers to mathematics, to the science of structures, to mathematical forms, and to numbers:

> The future of the theory of pictorial harmony is precisely here. The forms juxtaposed *in some way* are closely and deeply related. Their fundamental relationship will finally be able to be expressed in mathematical form, but in terms irregular rather than regular. Number is the ultimate abstract expression of all art. It is obvious that this objective element necessarily must have the cooperation of reason, of knowledge. It will be this, in the future as well, that will permit the artist to say not only *I was*, but *I am* [5].

On 4 April 1928 at the Friedrich Theatre in Dessau Kandinsky staged "*Pictures at an Exhibition*", the work for piano by Modesto Moussorgsky composed in 1874. Kandinsky was the scriptwriter and the set designer (Fig. 1). Silvana Sinisi, in her book written on the occasion of the 1983 reproduction of the performance by the Berlin *Hochschule der Künste*, said:

> *Paintings in an Exhibition* seems to indicate an attitude that is colder and more mental oriented towards a composition of geometric forms that show traces of the rigorous research methodology carried forth in the context of the Bauhaus. The relationship between the various elements shuns any narrative concatenation and purely external logic in order to follow criteria that are freer and articulated, where the laws of juxtaposition and contrast are established from time to time according to the dictates of an internal motivation. The *symbolic* form of the triangle recurs, arranged with the vertex on top, a large acute triangle divided into unequal sections with the smallest and most acute part turned

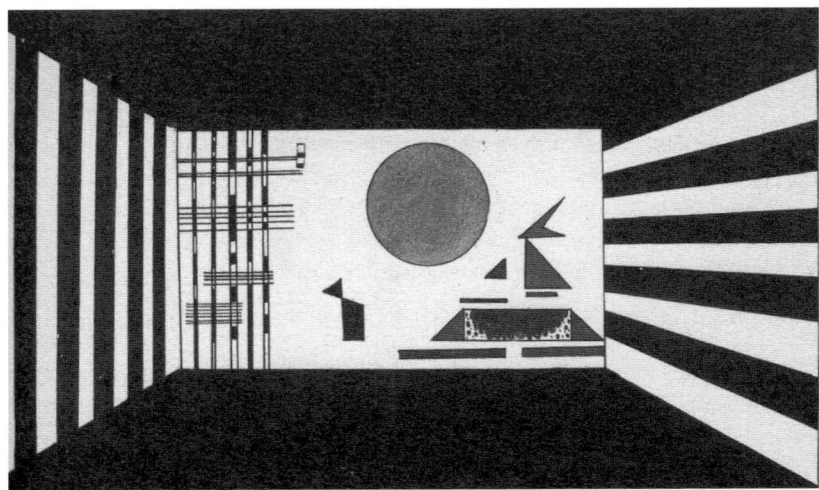

**Fig. 1.** Scene from *Pictures at an Exhibition* (see the section in colour)

towards the top: thus exactly representing spiritual life. Kandinsky tends to re-inforce the mathematical essence of the human body by encaging the bodily structure of the dancers in a geometric costume that limits and conditions the free expression of movement. The reduction of the human figure to primary geometric forms thus reveals the aim, like that of abstraction in painting, to grasp the essence of things and to realise in the specific context of the theatre a harmonic relationship between man and space based on a network of secret correspondences. The means used for this synthesis are plane geometric forms and pure colours that move, decompose and dissolve in space thanks to a know-ing play of gradations of luminosity obtained by transparency, direct lighting and projection with the help of lights worked by hand and reflectors [6].

It is worth noting how the great liberty in the use of space is connected to the use of "elementary forms" that are entirely Euclidean. But these forms, which have always been familiar to us and which have floated in our collective visual imagination for hundreds of years, become quite another thing on the canvasses of the artists of the 1900s. They are no longer "classics" that come down to us from ancient times of the Greek and Latin cultures, but are new "unknown" spaces that are explored for the first time, even if the "guides", almost as if they didn't want to get lost in the large field of abstraction, are "recognisable". One more proof that space doesn't exist in itself, but is a product of culture.

## Abstract/Concrete

In 1936 at the *Ecole Pratique des Hautes Etudes* in Paris, Alexandre Kojève gave a se-ries of lectures on Hegel, which were attended, among others, by Lacan, Queneau, Aron, Breton and Battaille. In one of the lectures, Kojève addressed the problem of

**Fig. 2.** Wassily Kandinsky, *White Cross* (*Weisses Kreuz*), January–June 1922. Oil on canvass, 100.5 × 110.6 cm. Peggy Guggenheim Collection, Venice. (Solomon R. Guggenheim Foundation, NY) (see the section in colour)

177

art, examining what he held to be one of its most mature manifestations: the art of his uncle Wassily Kandinsky. The text would be published with notes and additions by Kandinsky himself [7].

Kojève wrote:

> For centuries humanity has only known how to produce *representative* paintings, that is to say, subjective and abstract: every painting that embodies a Beauty already embodied in a real, non-artistic object is necessarily and essentially an abstract and subjective painting.

An abstraction of reality, the work of an artist who works subjectively. Instead, Kandinsky's art can be defined as true concrete art, given that

> … it is the art of embodying in a drawn painting a pictorial beauty that has never been embodied, is not embodied, and will never be embodied in any other place, not even in another real object other than in the painting itself, that is to say, in no real, non-artistic object. This art can be defined as the art of Kandinsky, since Kandinsky was the first to paint objective and concrete paintings starting in 1910.

Even more clearly, Kojève adds, if we take, for example, one of Kandinsky's drawings in which the artist embodies a Beauty implied by the combination of a triangle and a circle:

**Fig. 3.** Cover of the exhibit catalogue *Aux Origines de l'Abstraction*

178

This Beauty was not extracted or abstracted from a real, non-artistic object, which might even have been *beautiful*, but was a different thing. The Beauty of the painting *Circle-triangle* doesn't exist anywhere outside of this painting. In the same way the painting doesn't represent anything that is external to itself, its Beauty is purely immanent. It is the Beauty of the painting that exists only in the painting.

The conclusion is thus that if Beauty was neither extracted nor abstracted, but created from scratch, then it is not abstract but concrete: "the beauty of the painting is a Beauty that is real and concrete: the real painting is concrete". As the forms that appear in them are concrete, forms that Plato held to have always existed, and with him, many mathematicians. Perhaps here is where lies hidden the fascination of these very simple yet so evocative forms, so mysterious in the hands of a great artist in search of true "concrete" art?

Aristotle wrote that, in painting, if the most beautiful colours are applied without rules, less will be obtained than in realising a test palette of the subject itself.

### Mondrian

One of the most astonishing halls in the many museums of modern art can be found in the *Gemeentemuseum* in The Hague, Netherlands, where a series of Mondrian's canvasses are exhibited. Of trees, from a "realistic" tree (Kojève would say abstract) to trees that are increasingly rarefied, the various instances that lead to a geometric grid from which the object, that is, the tree, seems to have disappeared.

In 1920 Mondrian wrote:

> Neoplasticism has its roots in cubism. It can also be called abstract-real painting because the abstract (as in the mathematical sciences but without reaching their absolute) can be expressed by a plastic reality in the painting. That is a composition of coloured rectangular planes that expresses a more profound reality, which comes across by means of plastic expression of relationships and not by means of the natural appearance. It does what painting has always wanted to, but could only express in a hidden way. The coloured planes, as much by position as by their size and by the values of their colours, merely express relationships and not forms [...] The new plasticity poses its problems in aesthetic equilibrium and thus expresses the new harmony [8].

Theo van Doesburg, the other artist who, with Mondrian, founded *De Stijl*, added that the spiritual, the completely abstract, is precisely the expression of that which is human, since what can be sensed does not arrive at the height of the intellectual, and, in consequence, must be considered at belonging to a inferior level of human culture. "The universal basis of art is a balance of relationships and nothing else." It is in this sense that we speak of "stylisation" for the series of trees painted by Mondrian, or for the landscapes of Kandinsky of 1909–1910.

In 1910 Kandinsky executed the first abstract watercolour. According to Jacques Lassaigne:

> Gushing, the forms do not obey any construction that is imposed, conscious, premeditated; they do not follow a process of transposition, of knowing and spontaneous equivalences, but exist and are organised according to their natural inclination [9].

In 1914 Mondrian wrote:

> I construct lines and colour combinations on a flat surface, in order to express general beauty with the utmost awareness. Nature (or, that which I see) inspires me, puts me, as with any painter, in an emotional state so that an urge comes about to make something, but I want to come as close as possible to the truth and abstract everything from that, until I reach the foundation (still just an external foundation!) of things ...

> I believe it is possible that, through horizontal and vertical lines constructed with awareness, but not with calculation, led by high intuition, and brought to harmony and rhythm, these basic forms of beauty, supplemented if necessary by other direct lines or curves, can become a work of art, as strong as it is true [10].

**Fig. 4.** Piet Mondrian, *The Sea*, 1914. Charcoal and gouache on paper mounted on wood panel, paper 87.6 × 120.3 cm; panel 90.2 × 123.13 cm. Peggy Guggenheim Collection, Venice (Solomon R. Guggenheim Foundation, NY) (see the section in colour)

And again, in his famous book of 1935 entitled *Kn*, Carlo Belli writes of Kandinsky:

> The aspiration for a discipline came to Kandinsky during the war years. The ideas were pursued as far as they could be, and achieved as far as they could be. But this could not be done without order: thus the geometry is the starting point for these paintings as an ordering element. Kandinsky began to climb from 1922 on. Climbing up on the skeleton of a series of astonishing drawings he penetrates into the heart of the high mysteries of abstract equilibriums, and arrives at such a clarity and concision of form that they seem to spiritualise mathematics. Perhaps this was the dream of Pythagoras: the vertex of a spatial harmony that is heavenly, algebraic, absolute. The paintings of this period are dominated by a glacial and calculated geometric rigidity. From this vision is born a profound musicality that stays suspended in the cosmos by the mystery of a higher geometry. Order and measure: the supreme end of being [11].

### Mathematics and Art

We have mentioned cubism and neoplasticism, but we could also add futurism and suprematism [12]. One of the forms of ideal inspiration is the idea of space: its geometry, the rules of its mathematics, its logic.

It was the opinion of Max Bill in 1949 that,

It is objected that art has nothing to do with mathematics; that mathematics, besides being by its very nature as dry as dust and as unemotional, is a branch of speculative thought and as such in direct antithesis to those emotive values inherent in aesthetics … yet art plainly calls for both feeling and reasoning. … It is mankind's ability to reason which makes it possible to coordinate emotional values in such a way that what we call art ensues [13].

How then can mathematics be useful to an artist? Bill answers that,

Mathematics is not only one of the essential means of primary thought, and thus one of the necessary resorts for knowledge of the reality that surrounds us, but it is also, in its fundamental elements, a science of proportions, of behaviour from object to object, from group to group, from movement to movement. And since this science contains in itself these fundamental elements, and places them in meaningful relationships, it is natural that similar facts can be represented, transformed into images.

One last citation from El Lissitsky, taken from Lucy Adelman and Michael Compton's "Mathematics in Early Abstract Art" [14]:

He identified mathematics with the rational, with the conscious side and with the means of the artist.

Modern art and mathematics, a binomial that cannot be separated, at least at the beginning of the 1900s.

181

## Bibliography

[1] R. Osserman (1995) *Poetry of the Universe*, Weidenfeld & Nicholson.
[2] Wassily Kandinsky (1977) *Concerning the Spiritual in Art*. Trans. M. T. Sadler, New York, Dover Publications, p. 19.
[3] W. Kandinsky, *op.cit.*
[4] W. Kandinsky, *op.cit.*
[5] W. Kandinsky, *op. cit.*
[6] W. Kandinsky (1984) *Quadri di un' esposizione*, Editore Università Europea, Macerata, p. 65.
[7] A. Kojève, Kandinsky (2005) *Quodlibet*, Macerata, pp. 32–40.
[8] Mondrian, P. 1921. Le néo-plasticisme (principle général de l'équivalence plastique). *De Stijl*, February 1921: 18–19.
[9] J. Lassaigne (1964) *Kandinsky*, Skira, Geneva, p. 73.
[10] Reprinted in H. Janssen, J.M. Joosten (eds.) (2002) *Mondrian de 1892 à 1914. Les chemins de l'abstraction*, Edit. Réunion des Musées Nationaux, Paris, p. 196.
[11] C. Belli (1935) *Kn*, Edizione del Milione, Milan, p. 105
[12] See the paper "The Square: Homage to Bruno Munari" in this present volume. For a broader treatment, see M. Emmer (2007) *Visibili armonie: arte cinema teatro matematica*, Bollati Boringhieri, Torino.

[13]  M. Bill (1993) The Mathematical Way of Thinking in the Visual Art of Our Time, in: M. Emmer (ed.), *The Visual Mind*, MIT Press, pp. 5–9. This article was published the first time in German in 1949 in *Werk* 3, and was republished several times with the title "The Mathematical Approach in Contemporary Art". In the volume cited here it was revised by the author with some slight corrections.

[14]  L. Adelman, M. Compton (1980) Mathematics in Early Abstract Art, in: *Towards a New Art. Essays on the Background to Abstract Art 1910-20*, The Tate Gallery, London, pp. 64–89.

# The Language of Mondrian:
# Algorithmic and Axiomatic Investigations

Loe Feijs

**Abstract** In this article I describe a computer program that generates two dimensional divisions of the plane. The program is used as a personal tool to explore and help understand Mondrian's non-figurative paintings. Like the axiomatic method in mathematics, computer programming asks the designer to choose a minimal set of object types and to factor out commonalities in the principles of establishing relationships among the objects. Parallels are sought between the axiomatic approach, the computer program and Mondrian's quest for purity and non-figurative compositional principles.

## The Painter

The Dutch painter Piet Mondriaan was born in the Netherlands in 1872. He was trained as an art teacher and developed at first as a figurative painter in the tradition of, amongst others, The Hague school. Later he lived in Paris, London and eventually New York. He changed his name from Mondriaan into Mondrian. His first non-figurative works, related to Cubism, date from around 1912. His first steps on the road of abstraction were done through a process called "doorbeelding". Like Van Doesburg and Van der Lek, he transformed a figurative original over and over until eventually it became more or less non-figurative. The non-figurative abstract compositions made him famous. In the period 1910–1920 he restricted the topics: windmills, ocean coasts, trees, church towers, and flowers. Except for the flowers, these topics were subjected to experiments of further abstraction. He adopted a restricted set of colors, sometimes soft pastel colors, eventually only red, yellow, and blue, next to black, white and sometimes gray. He reduced the number of compositional elements till only horizontal and vertical directions were allowed. He removed most of the texture and shading from the color planes. Eventually this left him with a set of compositional elements that were pure and simple, yet spanned a rich search space to explore the essentials of composition. The most important of these elements are planes, lines and colors. The way Mondrian went from figuration to abstraction has been well-documented and has been the topic of exhibitions such as "From Figuration to Abstraction" (The Hague, 1988) and "The Path to Abstraction" Paris (2002), Fort Worth (2002), The Hague (2003), also see [1–3].

Mondrian's quest for pure beauty and balance, independent of figuration, did not stop. From 1915, when he had gone already all the way from figuration to abstraction, until 1944 when he died at the age of almost 72, he worked to find new ways of expression. If we call his quest *"from figuration to abstraction … and beyond"*, then the *"… and beyond"* part spanned a period of almost 30 years. As an admirer of Mondrian, and living in The Netherlands where I have easy access to both books and excellent museums, I got fascinated by this continued development of the abstract works.

## The Algorithmic Approach

Many attempts have been made to let computers generate Mondrian look-alike compositions. The first and most well-known was the work of Michael A. Noll, see [4,5]. For a survey of programs, see [6]. In the summer of 1993, I decided to try and create a computer program that generates two dimensional divisions of the plane. The program is meant as a personal tool to explore and help understand Mondrian's non-figurative paintings. Like the axiomatic method in mathematics, computer programming asks the designer to choose a minimal set of object types and to factor out commonalities in the principles of establishing relationships among the objects.

The program is in Turbo Pascal. It is 4500 lines of code. For each generated image the program performs the following steps: (1) set-up of a data structure and setting the positional and statistical parameters according to the composition type, (2) invoking a random generator to choose the specific Boolean and numeric properties of the objects such as their initial position, colour and certain so-called growth-parameters, and (3) developing the composition itself by letting the objects grow and letting them interact according to the chosen parameters.

In Fig. 1 this is illustrated for a division of the plane which resembles the "peripheral" type of composition from around 1922. Following the development from left to right, top to bottom, it can be seen that the random generator has produced 8 kernels, six of which are black, so-called A kernels, and two which are coloured, so-called B kernels. This is done in steps 1 and 2. Each kernel object has certain attributes which will determine its growth during step 3. The black kernels start growing in a very early phase. Note that some of them have the property of crossing, whereas others are non-crossing when they collide with something else. After that the coloured kernels, develop by growing into two dimensions. Only six snapshots are shown from a development which may actually take up to ten phases, each of ten growing-steps.

The following fragment describes the procedure ColorPlanes which performs steps (1) and (2). The first step is done by choosing the parameters A and B. Later

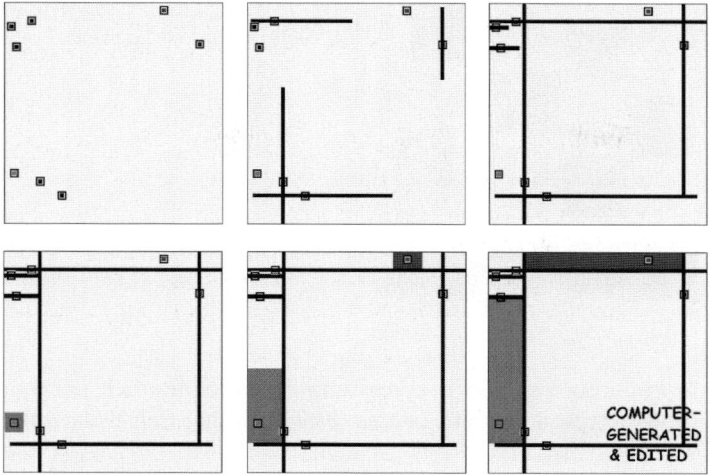

**Fig. 1.** Growth of lines and planes from random kernels (© Loe Feijs)

the A kernels will develop into early-phase growing black cells that appear as one-directional planes (hence become blacklines). The B kernels become cells will develop later into colored planes. Some of the properties are set in further subroutines such as mkKernelB. Normally, the B cells develop after the A cells, and the effect is that they "flood" the areas enclosed by the black lines (as shown in Fig. 1). The ColorPlanes type of composition happens to have only B cells. They stop growing when they collide with another B cell (actually their growth slows down already while approaching each other, so they usually keep some distance).

```
var
A,B,i : integer;
k : PKernel;
begin
  self.color := white;
  A := 0;
  B := 50 + random(50);
  MaxCell := A + B;
  for i := 0 to MaxCell do begin
    k := mkKernelB([rose,sky,gold]);
    with k^ do begin
      Maxlen := ((xMAX - xMIN) div 5);
      k^.ORagged := true;
    end;
    Cells[i] := mkCell(k);
  end; {for}
end;
```

185

**Fig. 2.** Computer-generated image of the *ColorPlanes*type (© Loe Feijs)

For more information about the program we refer to the article in Leonardo [6]. Figure 2 shows a typical outcome of step (3) when using *ColorPlanes* for steps (1) and (3).

For each composition type there is a procedure like this. Below the main case statement is shown: it embodies a simplified timeline and an overview of the main composition types that I tried to formalize.

```
case year of
    1914    : SymmetricPierOcean;
    1915    : AsymmetricPierOcean;
    1916    : ShortBlackLines;
    1917    : if random(2) = 0
                then ColorPlanesShortBlackLines
                else ColorPlanes;
    1919    : CheckerBoard;
    1920    : ColorPlanesLongBlackLines;
    1922    : Peripheral;
    1925    : ThreeSidedPeripheral;
    1928    : TwoSidedPeripheral;
    1930    : Cross;
    1932    : DoubledCross;
    1936    : MoreDoubledLines;
    1940    : CrossingLinesFewPlanes;
    1941    : Place;
    1942    : CrossingColoredLines;
end
```

The following figures show some of the variations of computer-generated images that can be obtained in this way.

It is interesting to see how I can model some characteristics reasonably well whereas other characteristics still resist formalization in the presented framework. For example the Place paintings like Place De La Concorde (1938–1943) and Trafalgar Square (1939–1943) were almost impossible to fit into the structure of the program. My formalization provides a new way to analyze and describe the syntax of

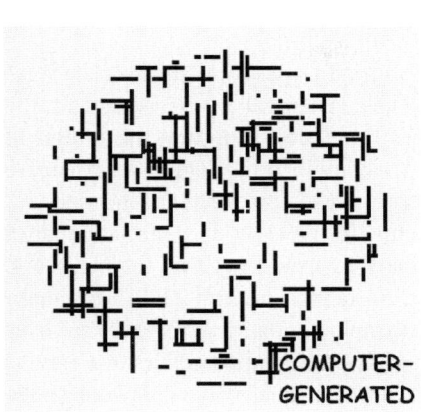

**Fig. 3.** Computer-generated image of the *ShortBlackLine* type (© Loe Feijs)

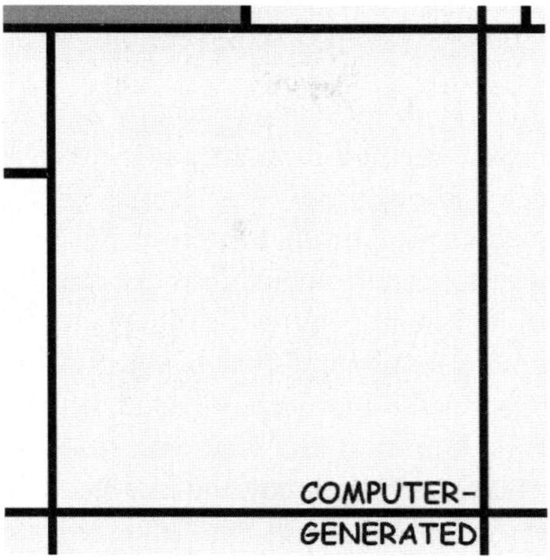

**Fig. 4.** Computer-generated image of the *Peripheral* type (© Loe Feijs)

187

**Fig. 5.** Computer-generated image of the *CrossingLinesFewPlanes* type (© Loe Feijs)

**Fig. 6.** Computer-generated image of the *CrossingColoredLines* type (© Loe Feijs)

some of Mondrian's works. Whenever my computer program produces an ugly or non-typical division of the plane, the question is "why doesn't this work?" Looking again at the real works of the intended type I usually find a compositional aspect easily overlooked otherwise.

# The Axiomatic Method

There is an interesting analogy between Mondrian's quest and the so-called axiomatic method in mathematics. An "axiom" is a statement that is assumed to be true without proof. The notion of axiom has developed alongside the notion of mathematical proof. In a proof, reliable facts are combined in a logical way to prove new statements, which thus are considered true as well. But what are the basic reliable facts to begin with? These are the axioms. In the early days of mathematics these were thought of as being so obvious that they would need no further proof. This was the approach of Euclid, who lived in Alexandria around 300BC. His main work, known as The Elements, stood as an example of good mathematics for over 2000 years. Also today, the idea that the axioms are self-evident occurs in a natural way to many students when they are introduced to the usual axioms of plane geometry for the first time. But the investigations of Bolyai (1802–1860) and Lobatchevsky (1792–1856) demonstrated that it is very well possible to develop another kind of geometry in which the fifth of Euclid's axioms simply does not hold. Klein (1849–1925) and Poincaré (1854–1912) created models of such alternative axiom systems and thus showed that the change in axioms did not introduce inconsistencies.

Actually, constructing proofs is only half of the story of geometry (or any other branch of mathematics conducted according to the axiomatic method). The other half of the story is about constructing the objects of study themselves. The story has remarkable parallels to the story of the axioms. In a geometrical construction, existing things are combined in a constructive way to make new things, which are thus considered to exist as well. But what are the basic object types to begin with? One can construct a tiling from triangles and rectangles; triangles and rectangles are constructed from points and lines; what are points and lines made of? These are postulated: it is assumed that points and lines exist a-priori, without any need for construction. Certain relationships may or may not exists between all these objects and that is precisely what the axioms and the proofs are about.

Let me come back to the explanation of the axiomatic method. It is easy to assume a lot of object types and powerful axioms and then construct an object having certain provable properties, but this is not always interesting. The art of mathematics according the axiomatic method is to work in a minimalist way. Start from very few object types and a minimal set of axioms and then explore the consequences, proving as much as possible, up to the limits of the system. So in summary, the mathematician looks for:

- a minimal set of object types;
- a minimal set of axioms;
- a maximal set of consequences,
- while minimizing the complexity of the proofs.

A good example is projective geometry. It has only two object types: points and lines. A given point may or may not be incident to a given line (incident means that the point is on the line, or which is the same, the line goes trough the point). The typical axioms say (1) any two lines intersect in a point, (2) any two points can be joined by

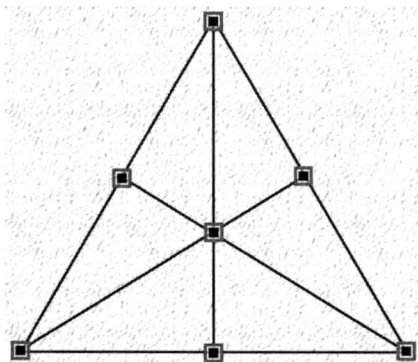

**Fig. 7.** Construction in projective geometry with seven points and seven lines (© Loe Feijs)

a line, (3) this point (or line) is unique if the given lines (or points) are distinct, (4) there exist at least a certain number of points and lines with certain relationships so as to make non-trivial constructions possible. We refer to Van Dalen [7] for further details about the axioms. Fig. 7 shows a construction built from seven points and seven lines (note that formally the "circle" is a line too, we didn't say anything about lines being straight). Each point lies on three lines. Each line goes through three points.

These insights in the nature of mathematics, which developed in the second half of the 19-th century gained additional interest, both from scientists and non-scientists through the development of relativity theory. First the special theory of relativity by Einstein (1879–1955) that was published in 1905. Thanks to an observation of Minkowski (1864–1909) this theory could be explained in terms of a four-dimensional non-Euclidian space. Then in 1915 Einstein's general theory of relativity appeared, which states that this space is curved. Although the details were far too difficult and advanced for non-mathematicians, the entire terminology had a great appeal to art movements such as Futurism, Cubism, Constructivism and Suprematism. De Stijl was no exception.

In 1923 an article written by the mathematician Poincaré was reprinted in De Stijl, the journal edited by Van Doesburg [8]. The article is called "Pourquoi l'espace à trois dimensions?" This article discusses the possibility of alternative geometries, very much in the line of reasoning developed by the mathematician Felix Klein. It is argued that there is a geometry called Analysis Situs (Topology) which abstracts away from metric properties of traditional geometry. It is preceded by a heading in Dutch: De beteekenis der $4^e$ dimensie voor de nieuwe beelding (The relevance of the 4th dimension for the new plastic art). This article was also presented at the Matematica e Cultura 2005 conference by Michele Emmer, who I thank for bringing it to my attention. In my view it is very likely that Mondrian has seen this article, although I think it is unlikely that he was fully aware of the mathematical technicalities behind Poincaré's article.

The analogy between Mondrian's quest and the axiomatic method in mathematics which I tentatively propose, is that Mondrian has been looking for:

189

- a minimal set of object types;
- a minimal set of compositional rules;
- such as to maximize harmony and balance;
- while minimizing the re-introduction of figuration.

The computer program described in this article shows that many composition types can be reconstructed from two object types: lines and planes with colours from a very restricted set. The composition principles are modelled through the algorithmic part of the program, which, for a given composition type, executes in three steps: (1) set-up of a data structure and setting the positional and statistical parameters according to the composition type, (2) invoking a random generator to choose the specific Boolean and numeric properties of the objects such as their initial position, colour and certain so-called growth-parameters, and (3) developing the composition itself by letting the objects grow and letting them interact according to the chosen parameters.

I do not claim that the computer-generated images do have the same (or even approximately the same) aesthetic qualities of the real Mondrian; on the contrary, the longer I tried to refine the computer program, the more my appreciation for the real works grew. But while working on the program, I learned to look at Mondrian's non-figurative work in another way. Designing the program forced me to develop a very precise vocabulary, embodied in the names of the programs variables, the record fields, and the definitions of the various algorithmic procedures involved.

An example of a "lesson" thus learned is the following: in the program, there is no formal distinction between lines and planes. A line develops differently during the growth process (it grows only in the X or the Y direction, whereas a plane grows in two directions). This can be viewed as an example of minimising the set of object types.

## Avoiding the Re-Introduction of Figuration

The program deals with the object types and the composition principles. Then I got the book of Carel Blotkamp [9] which got me even more enthusiastic. The book gives a very balanced survey of Mondrian's work, but I got particularly triggered by the analysis of what Blotkamp calls "the Art of Destruction". This helped me a lot to look at a more subtle way to certain compositional principles. Many of which were not easy to encode into the program, but nevertheless (or better: precisely for that reason) I found them interesting. In this section, I take the opportunity to discuss three of those principles and demonstrate them using some generated bitmaps.

In my own words, "*the Art of Destruction*" is about the idea that Mondrian wanted his compositions to have an aesthetic quality of their own, without reference to figuration. He sought to hide the object-character (individuality) of the compositional elements and avoid those interactions between elements which reintroduce three-dimensional effects. For example, gray or black lines separating the color planes serve to avoid the impression that the color planes float in front of a white background, which is a "problem" with e. g. Composition with Color Planes 2, 1917.

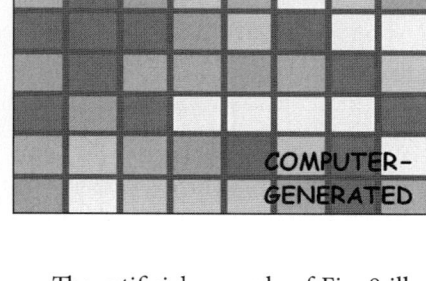

**Fig. 8.** Seeing floating planes (*top*) and a "solution" (*bottom*) (© Loe Feijs)

The artificial example of Fig. 8 illustrates the first effect. Indeed, the composition principle of the "checkerboard" does eliminate the effect of depth. But there is a price to be paid for this innovation: this set of object types seems too minimal; instead of yielding a rich variety of compositions, the approach gets exhausted soon. Mondrian soon sought to find ways to soften the rigidness of this type of grid (see the dicussion and Mondrian's correspondence with Van Doesburg in Bois et al. [2]). Gregory Schufreider [10] explains the abolition of the checkerboard grid in favour of a more flexible grid as using the grid as a space opening device, a "structure of openness".

As another example of howto hide the object-character of the compositional elements, the doubling of lines serves to "destruct" the individual form of lines. This is illustrated in Fig. 9. The left computer-generated image is an approximation of a certain type of compositions made around 1930–1932. The right computer-generated image shows two doubled lines and a way of employing them that are typical for the period 1932–1935.

A second example concerns ways of introducing some ambiguity into the definition of a plane. This is illustrated in Fig. 10. In the left computer-generated image, there is one colored plane object. In the middle image one plane is surrounded by black lines, but due to the way one of those lines does not precisely fit the plane, the object-character and the rigidness of the plane is somewhat weakened. The doubling of lines serves to "destruct" the individual form of lines. In the right computer-generated image, four adjacent planes are at the same time one super-plane, again somewhat weakening the definition of the plane. For the art-historically correct

191

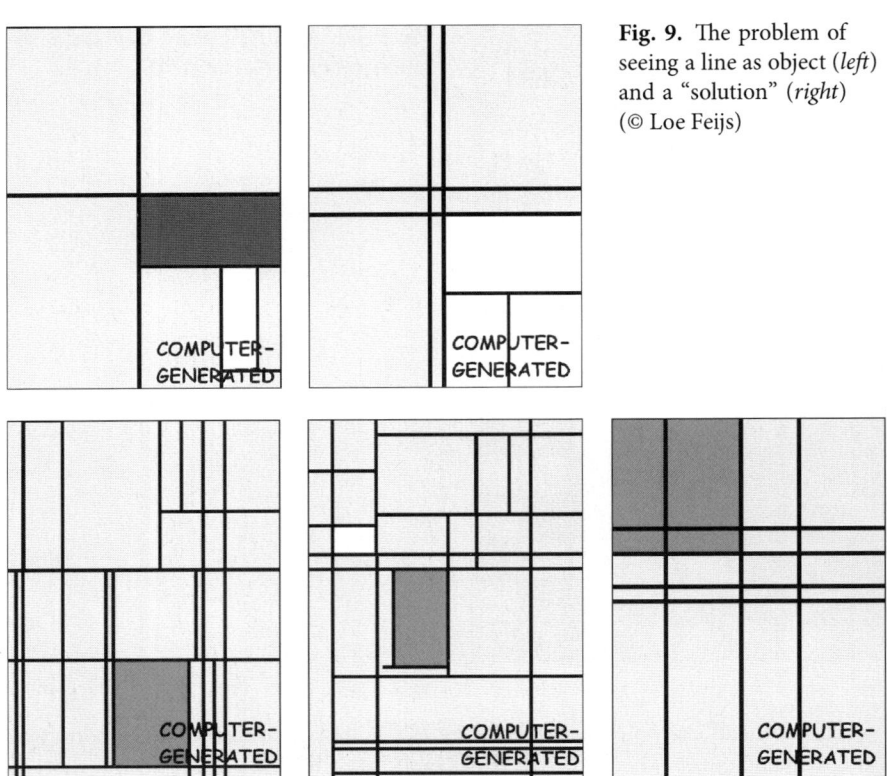

**Fig. 9.** The problem of seeing a line as object (*left*) and a "solution" (*right*) (© Loe Feijs)

**Fig. 10.** Seeing a plane as object (*left*) and two "solutions" (*middle* and *right*) (© Loe Feijs)

terminology and other, more subtle examples in the real paintings, I refer to [9]. Here I use a kind of engineering terminology of problems and solutions. Regarding Fig. 10 the first solution was already found around 1920 whereas the second appeared in 1936.

Comparing the "destruction" theory of Blotkamp's book [9] to the programming approach, the former is analytical whereas the latter is synthetic (generative). But there is another difference: Blotkamp's explanation includes knowledge about human perception, for example the fact that people "see" three-dimensional effects. My computer program is only of a syntactic nature and its only built-in knowledge is two-dimensional. This is an advantage when developing or (later, perhaps) refining the program; it forces me to study and describe composition types in a very syntactic way. As such, my approach has no power of explanation - in contrast to Blotkamp. The distinction is important since it demonstrates that there are at least three possible levels of describing composition types:

1. syntactic (lines and planes),
2. perceptive (seeing or not seeing lines/planes or 3D effects), and
3. semantic (seeing or not seeing trees, churches, the sea, for example).

**Fig. 11.** Three levels of describing composition types (© Loe Feijs)

My approach helps to understand that the first two levels do not coincide. In terms of this three-level model, Blotkamp can be said to show that Mondrian's non-figuration is not just the absence of semantics, but also the absence of options for perceptive interpretation. The situation is visualized in Fig. 11. The crossed arrows mean "to be avoided".

Now we can re-phrase the idea of avoiding the re-introduction of figuration in a much more precise way as finding constructions that:

- do not represent reality, and
- do not even have real-world perceptual qualities.

Just to show the explanatory power of this reasoning, let me apply it to the compositions of the *CrossingColoredLines* type, like the computer-generated Fig. 6 or the real New York City III (uncompleted) 1941–1942. Here the search for a minimal set of object types has been pushed to an extreme: lines and planes have become unified into one object type (the coloured line). But this endangered the last optimisation condition (minimizing the re-introduction of figuration). Although the compositions do not represent reality (no landscape, windmill, church etc.), there is a real-world perceptual quality where the lines cross. At each crossing one of the lines appears to run over the other. The observer perceives depth. This has been solved in Broadway Boogie Woogie (1942–1943) where all lines are yellow and where each crossing is covered by a small red, blue or grey square (modelling the Boogie Woogies was outside the scope of my programming effort). In the Boogie Woogies there is another innovation in the search for a minimal set of object types: the concepts of planes and the canvas (the whole composition) are unified. Certain planes contain a kind of sub-compositions.

193

## Concluding Remarks

A number following conclusions have already been mentioned in [6], such as that the computer modeling of types of divisions of the plane is a useful tool for observing and classifying Mondrian's abstract works. This leads to a typology which is of a syntactic nature (not including perceptive or semantic interpretation). In this article we explored the analogies with the axiomatic approach in mathematics.

Like the axiomatic method in mathematics, computer programming asks the designer to choose a minimal set of object types and to factor out commonalities in the principles of establishing relationships among the objects. Mondrian's investigations from 1915 onwards can be interpreted as a search for an "axiomatization" in the sense of looking for a minimal set of object types, a minimal set of compositional rules such as to maximize harmony and balance while minimizing the re-introduction of figuration.

## Acknowledgements and Credits

The copyright of Mondrian's works belongs to Holzman Trust, c/o HCR International. The computer program described is not available, nor will license be granted to reproduce the computer generated images. I would like to thank Jacqueline Cove, Jan van der Lubbe, Carel Blotkamp, Kees Overbeeke and Michele Emmer for their help and support for the research described in this article.

## Bibliography

[1] Herbert Henkels (red.). *Mondrian. From Figuration to Abstraction*, Tokyo, Shimbun. 1987

[2] Y.A. Bois, J. Joosten, A. Zander Rudenstine, H. Janssen. *Piet Mondriaan 1872–1944*, Milan, Leonardo Arte, 1994. Catalogus voor tentoonstellingen in Washington, Den Haag en New York in 1995 en 1996.

[3] J.J. Joosten, R.P. Welsh. *Piet Mondrian: Catalogue Raisonné*. New York: Harry N. Abrams (1998).

[4] A.Michael Noll. *Human or machine, a subjective comparison of Piet Mondrian's Composition with Lines and a computer-generated picture*, The psychological Record Vol 16 No 1, pp. 1–10 (1966), reprinted in: James Hogg (Ed.), Psychology and the visual arts (London: Penguin Books), pp. 302–314 (1969).

[5] A.Michael Noll. *The beginnings of computer art in the United States: a memoir*, Leonardo, Vol 27, No 1, pp. 39–44 (1994).

[6] Feijs, L. M. G. (Loe M. G.) *"Divisions of the Plane by Computer: Another Way of Looking at Mondrian's Nonfigurative Compositions"*. Leonardo – Volume 37, Number 3, June 2004, pp. 217–222, The MIT Press.

[7] D. Van Dalen. *Logic and structure* p. 87, Springer Verlag (1980).

[8] Henri Poincaré, *"Pourquoi l'espace à trois dimensions?"* De Stijl VI, 5, p. 66. See http://www.fi.uu.nl/~aad/documents/Utopie4Dprintversie.pdf for an online version.

[9] C. Blotkamp: *Mondrian: The Art of Destruction* London: Reaktion Books); ISBN: 1861891008 (1994).

[10] G. Schufreider. *Mondrian's opening: the space of painting*, Lecture delivered in the Department of Fine Arts, Harvard University, http://www.focusing.org/schuf.html, April (1997).

# Nature-Mathematics. An Operative Language*

Victor Simonetti

My life is constantly stirred by how incredibly surprising existence is. Since it is impossible to be indifferent to the beauty of nature, I am driven to search for its language. This research into nature has led me mainly to observe of the processes of its organisation, the better to capture its beauty and its mathematical character. I think that this language serves it as a base, used with the same liberty as a written language, and can be used with poetic creativity in the communication of messages and in the realisation of things in a way that is closer to our own nature.

We have only one model of nature, interpreted by our senses and intelligence, which is surely different from the physical reality. From our Flatland, with our limitations, we scientifically derive only relationships, letting the essence escape us. To be sure, the discipline par excellence of relationships is mathematics. Today we know that nature is made up of systems, systems of systems, interrelated in various and continually changing ways (as in becoming). Beauty encompasses, besides the formal aspect, the entirety, at every level, including the process. I wonder if in the relationship between nature and beauty, nature and mathematics, and mathematics and beauty, is there is not such a thing as "aesthetimatica".

Every system, every organisation and every relationship comes to life by means of actions. Every action, from the most elementary to the most complex, is an event that is inseparable from space-time using energy. An action has the nature of a vector: a point of application, direction, orientation, and intensity. It produces a work. It is the basis of everything: of every system and every existence that we know of, of every language. But languages are systems and systems theory is a field of mathematics.[1]

* *Translated by Kim Williams*

[1] *Personally the use of the concept of the vector has turned out to be an excellent way to interpret the natural organisations and even some of the aspects of human behaviour in architectural environments. I have used them methodically to create my works, especially with the use of physical forces, such as, for example, in the series: Moduli gravitazionali (Gravitational modules), Moduli energetici (Energetic modules), Espressione energetica (Energetic expression), Disgregazione armonica (Harmonic disintegrations), Flussi (Flows) and others. In Flussi the user has to participate by moving the work in order to read it. This path has led me to work directly with mathematics and the images refer to it.*

The Pythagorean series (from 1989) was born from plastically developing the Pythagorean theorem, considering it a natural theorem. With this series begins the use of mathematics as a language of nature, with the main aim of communicating messages that are aesthetic and didactic. The Fibopythagorean works (from 1995) conjoin in the process of their formation the Pythagorean theorem with the Fibonacci series, which is applied here in an unusual way to the area of squares.

## Pythagorean Works

The Pythagorean theorem also surprised me because of the immense number of new forms of organisation that it could develop, and by using it with a certain awareness, because of the nearness of the images to organic ones (that is to say, inversely, from mathematics to nature).

My reading of beauty in the organisation of nature has put me in contact with and has led me to use with a great degree of freedom the theory of vectors, sets, graphs and grids, flows, with the various geometries, algorithms and concepts that are indispensable for knowing new systems as well. Working with mathematics to obtain new organisations is like working with nature. I cannot forget dear Bruno Munari, with whom I shared this discussion, who made me publish my first *Ciao Pitagora* (Ed. Corraini, and developed and continued by the publishing house Simonsegni, *www.simonsegni.it*) and who also set for me the example and inspiration that are still the driving forces for continuing the research.

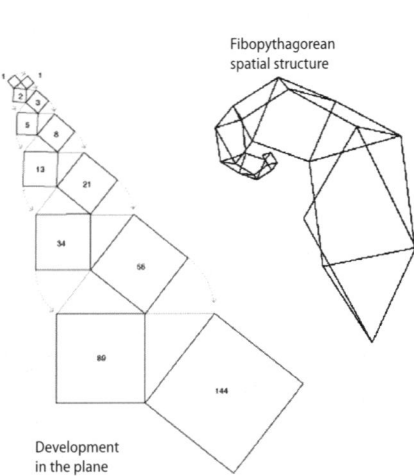

**Fig. 1.** The relationship between the Fibonacci series and the Pythagorean theorem

**Fig. 2.** Fibopythagorean spatial structure, 1998

**Fig. 3.** Fibopythagorean plane structure, 1995

**Fig. 4.** Contracted fibopythagorean structure, 1995

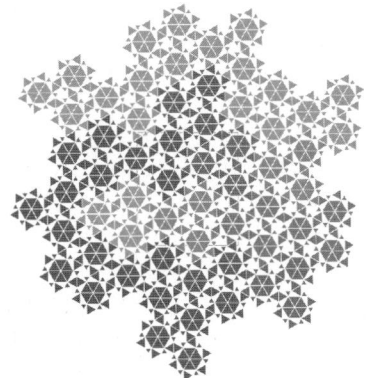

**Fig. 5.** Pythagorean constellation, 2004

**Fig. 6.** Version of the Pythagorean tree, 2004

197

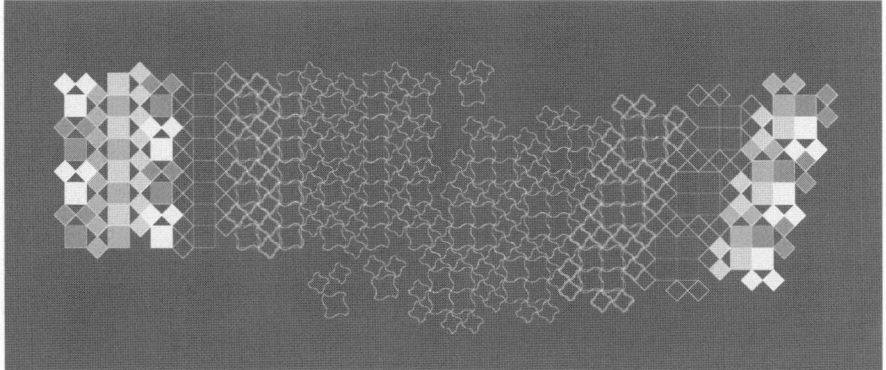

**Fig. 7.** Pythagorean metamorphosis, 2004

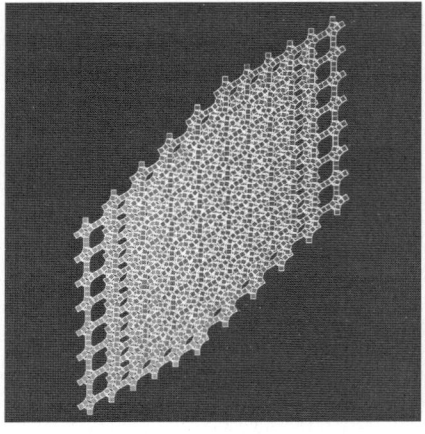

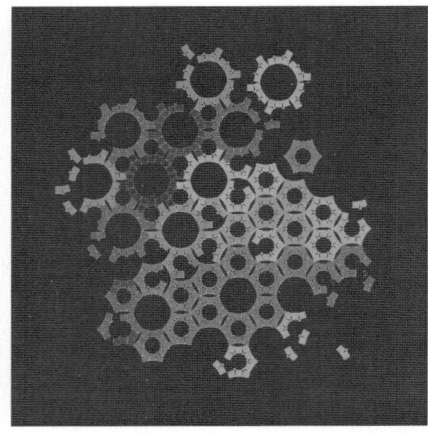

**Fig. 7.** Complex of Pythagorean grids, 2004

**Fig. 8.** Structure of Pythagorean mole-cules, 2004

# Mathematics and Words

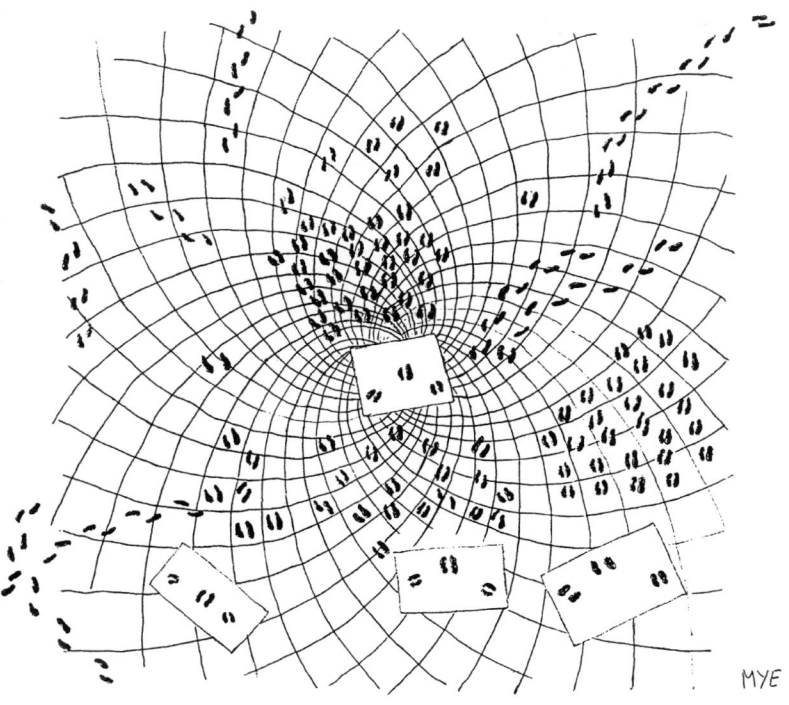

MYE

# Prime Time Entertainment

Marcus du Sautoy

When David Beckham moved to Real Madrid, there was a lot of speculation in the British media about why he chose the 23 shirt. Some suggested it was a cynical move by Real Madrid to sell a lot of football shirts in America. Americans don't know much about football – basketball and baseball are more their sports. The most famous basketball player in America was Michael Jordan who played in the number 23 shirt. The British media suggested that Real Madrid were trying to trick Americans into buying Beckham's replica shirt just because it had Michael Jordan's 23 on the back.

Other pundits suggested the choice of 23 was much more sinister. History warns that 23 might be a bad number to put on your back – Julius Caesar was assassinated by being stabbed 23 times in his back by Roman senators. More wild suggestions ranged from connections with Star Wars movies – Princess Leia was imprisoned in cell AA23 – to more scientific theories – there are 23 chromosomes in our bodies.

Then one radio station noticed that I'd just published a book "The Music of the Primes" whose UK hardback cover features an Arsenal 23 football shirt. I was immediately whisked into the studio to give my explanation of the significance of Beckham's choice. The cover of the book I wrote not only featured the 23 football shirt, but also my front door – number 53 – my local bus in London – number 73 – and a mystery novel open at chapter 13. The connection between all the numbers were that they are all prime numbers, numbers that are only divisible by themselves and 1.

Prime numbers are the most fundamental numbers in mathematics. They are the atoms of arithmetic. Every number is built by multiplying prime numbers together: for example $105 = 3 \times 5 \times 7$. In chemistry, the Periodic Table lists all the atoms from which matter is made. For the mathematician, primes are like the hydrogen, helium and lithium of the world of numbers. They are the building blocks of our subject.

On the way to the studio to talk about Beckham's choice to play in a prime, I began to realise it wasn't just a coincidence. As I looked through Real Madrid's team, all the key building blocks of their squad all play in prime number shirts! Carlos at number 3, Zidane at number 5, Raul at number 7, Ronaldo at the time played in the 11 shirt. It is clear that Real Madrid have recognized the importance of these indivisible numbers. Each new building block in the team gets a prime number shirt.

Anton Berg like David Beckham also played musically in the number 23 shirt. In the *Lyric Suite* a 23 note sequence is used to represent Berg which he intertwines with a 10 note sequence meant to represent his lover. Interestingly the Ancient Chinese used to believe that even numbers were female numbers and odd numbers male. But of all the odd numbers, it was the prime 17 and 23 who were the macho numbers because they couldn't be broken down or factored. Odd numbers that were divisible like 15 or 21, had in Chinese culture a rather effeminate character.

John Nash, the Nobel winning laureate made famous by the recent movie *A Beautiful Mind*, also took rather a fancy to the number 23. Indeed one of the first signs of his impending madness was an occasion when he went round the department trying to convince people that the picture of Pope John Paul XXIII on the front cover of Life magazine was in fact Nash in disguise. His argument hinged on the importance of 23.

Given the striking effect that playing in primes has had for the success of Real Madrid, I recently persuaded the team I play for to buy a new strip all with primes on the back. We now play in shirts from 2 to 41. In our old strip we had finished the previous season at the bottom of the Super Sunday League Division 2. The effect of the primes was extraordinary. We shot to the top of the table and got promoted to Division 1. You can see our football team in a 4 minute movie on the web (http://www.spiked-online.com/sections/science/sciencesurvey/films.stm) explaining the first great theorem of mathematics: Euclid's proof that, however big our team gets, there will always be another prime number football shirt out there for our next great signing.

Rather than trying to emulate Beckham and Berg, I play in the number 17 shirt – a rather special prime in fact. Prime numbers of the form $2^n + 1$ are called Fermat primes. If you take a regular shape with a Fermat prime number of sides then there is a special way to reproduce the figure using just a straight line and a compass. The famous mathematician Carl Friedrich Gauss discovered the construction for the 17-gon and was so excited by the beauty of the geometry that he wished for the construction to be inscribed on his grave-stone. Its discovery was one of the principal inspirations for Gauss choosing a career in mathematics.

The number 17 is also the favourite prime of a very strange species of cicada that lives in the forests in North America. Called Magicicada septendecim, the species has a very curious life cycle. The cicadas hide underground doing nothing for 17 years except gnawing away at the roots of the trees. Then after 17 years the cicadas emerge simultaneously en masse into the forest for a six week party. The sound of the cicadas is so deafening that residents in the area are forced to move out. The cicadas party away, eating the leaves, mating, laying eggs for the next generation and then after six weeks, the cicadas all die and the forest goes quiet again for another 17 years.

Is it just a coincidence that the cicadas have chosen a prime number of years to hibernate underground? It seems not. There is another species which hides underground for 13 years and another for 7 years. So what is it that is so special about the primes for these cicadas? Scientists have a theory that there used to be a predator that also appeared periodically in the forest. The predator liked to crash the cicadas' party and eat all the cicadas. The cicadas discovered that by choosing a prime number of years to hide underground they had the best chance to keep out of synch with the predator. For example, if the predator appears every 6 years then the cicada who appears every 7 years stands a better choice of surviving than those who appear every 8 or every 9 years. They also appear more often in the forest.

It seems that a competition evolved between the cicada and the predator to see who could find the next prime number. The cicadas turned out to know their primes better than the predator reaching as far as 17 in certain forests. The predator has subsequently died out unable to emerge in synch with the cicada. The primes are the key to the evolutionary survival for this species of cicada.

The strategy used by the cicadas to keep cycles out of synch is also employed by the composer Olivier Messiaen. Messiaen was obsessed with prime numbers and uses them in the opening movement of *The Quartet for the End of Time*. The quartet was composed whilst Messiaen was in a concentration camp during the Second World War. He wrote the piece for the strange collection of instruments that he had available to him in the camp: a clarinet, violin, cello and piano. Messiaen uses the primes to create a sense of timelessness in the piece. The first movement contains a 17 note theme played alongside a 29 note theme. Because both numbers are prime they keep meshing in different ways as each theme begins again. It is not until you have heard the 17 note sequence 29 times that you hear the two themes repeating themselves together again. The primes for Messiaen help create a music with a sense of no end.

There is an art installation in London which plays on the same theme. Developed by Jem Finer, once of the Pogues, Longplay is a sound installation which won't repeat itself until December 31 2999. By using six different themes with coprime durations, the piece exploits the same trick as Messiaen's quartet.

Just as the cicadas have found that 17 and 13 are life-saving numbers, primes are also the key to the survival of six characters in the movie *The Cube*. The movie is set in a huge maze which resembles a complex Rubix cube. Each room in the maze is a cube shaped room with six doors leading through to more rooms in the maze. Six characters wake up to find themselves inside this maze. They have no idea how they got there but they have to find a way out of the maze. The trouble is that some of the rooms are booby-trapped. The protagonists need to find some way to tell before they enter a room whether it is trapped or not – or else a whole array of horrific deaths await them including being incinerated, covered in acid or being cheese-wired into tiny cubes.

One of the characters is meant to be a mathematical whiz at school. She suddenly makes the breakthrough that the numbers at the entrance to each room hold the key to revealing whether a trap lies ahead or not. It seems that if any of the numbers at the entrance are prime, then the room contains a trap. "You beautiful brain" declares the leader of the group at this piece of mathematical deduction. Actually it turns out that they also have to watch out for prime powers but this proves beyond the clever school kid. Instead they have to rely on an autistic savent who is part of the group. In fact he turns out to be the only one to make it out of this prime number maze. Anyone who is having trouble getting their kids to learn their multiplication tables might find the film a great piece of propaganda for getting them to know their primes.

It is not only the movies that have used primes as a voice for autistic savants. The best-selling novel *The Curious Incident of the Dog in the Night-time* by Mark Haddon tells the story of the aspergic Christopher. He likes the mathematical world because he can understand how it will behave. The logic of this world means there are no surprises. It is a secure domain, the cupboard under the stairs where Christopher can retreat to. Human interactions in contrast are full of uncertainties and illogical twists that Christopher cannot cope with. His story is told through a sequence of chapters that are numbered by the prime numbers. As Christopher explains "I like prime numbers … I think prime numbers are like life. They are very logical but you could never work out the rules, even it you spent all your life time thinking about them." The primes for Christopher combine an interesting mix of the logical mathematical world and the uncertain human world.

In *The Man who Mistook His Wife for a Hat*, Oliver Sacks also describes the case of two real life autistic savants who used the primes as a secret language. The twin brothers would sit in Sacks's clinic swapping large numbers between themselves. At first Sacks was rather mystified by their dialogue. But that night he cracked the secret to their code. Mocking up on some prime numbers of his own, he decided to test his theory. The next day he joined the twins, sitting quietly as they exchanged six digit numbers. After a while Sacks took advantage of pause in their prime number patter and added a seven digit prime. The twins were taken somewhat by surprise by Sacks's contribution. They sat thinking for a while. This was after all stretching the limit of the primes they had been exchanging to date. But then both twins smiled simultaneously, as if recognizing a friend.

During their time with Sacks, they apparently managed to reach primes with 9 digits. Of course no one would be too surprised if they were simply exchanging odd numbers or square numbers. But the striking thing about their achievement is that the primes are so randomly scattered amongst the numbers that their ability to find them is truly remarkable. Some have speculated how they did it. One explanation relates to another feat that the twins were adept at. The twins would often appear on television and impress audiences with their ability to identify quickly that October 23 1901 was a Wednesday. The ability to name the day of the week from a date relates to working with modular or clock arithmetic.

It maybe that the twins discovered that modular arithmetic is also the key to a method to quickly identify whether a number is prime. If you take a number like 17 and you calculate $2^{17}$ modulo 17 and the answer is 2 then that is a good piece of evidence for the number 17 being prime. Fermat proved that if it isn't 2 then that certainly implies 17 is not prime. So in general if you want to test if p is not a prime, then calculate $2^p$ and if the answer isn't 2 modulo $p$, then $p$ can't be prime. This test for primality is often wrongly attributed to the ancient Chinese. It only appeared for the first time in the seventeenth century. Some have speculated that given the twins aptitude to identify days of the week, they may well have been using this modular arithmetic test to find primes.

At first, mathematicians thought that if $2^p$ does turn out to be 2 modulo $p$ then that meant $p$ is prime. However it turns out that it is not a test that will guarantee primality. $341 = 31 \times 11$ is not prime yet $2^{341}$ is 2 modulo 341. But this example was not discovered until 1819. It is possible that the twins might have been aware of a more sophisticated test that would weadle out 341. Fermat actually showed that the test can be extended not only to powers of 2. Fermat proved that if $p$ is prime then for any number $n < p$, $n^p = n$ (modulo $p$). So if you find any number $n$ for which this fails you can throw $p$ out as a prime impostor.

For example $3^{341}$ is not 3 modulo 341 but 168. It isn't possible that the twins were checking through all numbers less than their candidate prime. Although modulo arithmetic is still easier that testing divisibility that is still too many tests for the twins to check through. But Erdös actually estimated (although did not rigorously prove) that to test if a number less than $10^{150}$ is prime or not, finding one time on the clock which passes Fermat's test already means that the odds that the number is not in fact prime are as little as 1 in $10^{43}$. So for the twins, probably one test was enough to give them the buzz of prime discovery.

It is not only Sacks' autistic twins that have been using primes for communication. Science fiction writers often choose mathematics as the way their aliens communicate with earth. Don Delillo's early novel *Ratner's Star* explores the idea of a mathematical message sent from outer space. In Carl Sagan's novel *Contact*, the heroine of the book Ellie Arroway works for SETI, the Search for Extra Terrestrial Intelligence. After months of listening to the background crackle of the universe she suddenly picks up a message which she realises is not white noise but a sequence of numbers. As she analyses the numbers she realises they are all prime numbers. As Jodie Foster says in the film version of Sagan's novel: "holy shit, there's no way that can be a natural phenomenon". Indeed it is an alien culture trying to get our attention by using the primes to say hello.

What is interesting about Sagan's choice of the primes is that it expresses a belief held deep by most mathematicians. Prime numbers more than any other part of our mathematical heritage have a timeless, universal character. Prime numbers would be there regardless of whether we had evolved sufficiently to recognize them. As the Cambridge mathematician G.H. Hardy said in his famous book *A Mathematician's Apology*: "317 is a prime not because we think so, or because our minds are shaped in one way or another, but *because it is so*, because mathematical reality is built that way".

That is why we too use primes in our attempts to communicate with any intelligent life out there. Many of the space ships launched by NASA have a tablet containing some attempt to talk to another culture should the ship eventually crash land on an inhabited planet. NASA often choose pixelated pictures where the number of pixels chosen is a prime number.

In *Contact* the primes are used by the aliens as a chat-up line. The same trick is used by Jeff Bridges in the film *A Mirror has Two Faces*. Bridges plays a mathematics professor and explains one of the great prime number mysteries in his first date with the English Literature professor, played by Barbara Streisand: The Twin Primes Conjecture. Once again this film illustrates one of the great sterotypes of the movie industry: mathematician equals socially inept, unworldly thinker who finds it impossible to engage with the physical, in this case sex. As Bridges and Streisand sit drinking their aperitif, he explains how he doesn't like to dance but prefers to watch:

Bridges:   "pairs ... it's interesting how coupling appears all through nature ... in mathematics ..."

Streisand:  "You were telling me something about primes."

Bridges:   "Yes, the Twin Prime Conjecture. It explores pairs of prime numbers, like 11, 13 or 17, 19. What was discovered was that it often occurred that primes were separated by ..."

Streisand:  "... one number in between."

Bridges:   "Exactly, exactly ... did you read my book. This is really marvellous."

Streisand:  "First date that I feel like I'm winning on a game show."

Bridges:   "Sorry. It's just so rare that I meet a person that I can discuss these things with."

Streisand:  "This Twin Prime Conjecture is interesting. What would happen if you counted passed a million? Would there still be pairs like that?"

At this point Bridges almost falls off his chair with excitement:

Bridges:   "I can't believe you thought of that. This is exactly what has yet to be proven in the Twin Prime Conjecture."

The Conjecture says that you will always find clusters of primes where $N$ and $N + 2$ are both prime. The first twin that Streisand would find beyond a million is the pair of primes 1,000,037 and 1,000,039. Unlike Euclid's beautifully simple proof that the primes go on for ever, no one has yet come up with an argument to show that infinitely often you will see these twin primes.

Communicating using primes is now much more pervasive than the occasional chat up line. Every time computers talk to each other and want to exchange confidential information, they are using a cryto-system called RSA which exploits the power of primes to keep messages secure. Every time a consumer visits a website and wants to send the site her credit card details she receives the website's crypto-number which her computer then uses to encrypt her credit card. The crypto-number is not a prime but the product of two prime numbers. The beauty of these prime number codes is that to *decrypt* the credit card, the website needs to use the primes which built the website's crypto-number. So for a hacker to crack the codes on the internet it suffices to find a method to find the primes which build large numbers.

The chemists have this wonderful machine called a spectrometer which if you give it a molecule the machine will tell you the atoms that built that molecule. For example give it salt and it says the molecule is made of sodium and chlorine. If one thinks of the analogy of primes as atoms like sodium and chlorine, then a hacker needs to invent some sort of prime number spectrometer. The trouble is that mathematicians have not yet understood the primes well enough to come up with a smart efficient way to crack a number with two hundred digits into its prime divisors.

The devastating effect of the discovery of such a method is illustrated in the films *Sneakers*. Doctor Janek a mathematician has made a breakthrough "of Gaussian proportions" as he describes it: he's found a way to crack the prime number codes on the Internet. He quickly gets murdered (not a good advert for becoming a mathematician!) and the mathematical code-cracker falls into the evil hands of Ben Kingsley. Robert Redford and his team save the day. But the movie ends with a news bulletin announcing that the Republican National Committee is bankrupt. "Apparently they had plenty of money in their accounts last week but today they just don't know where the money has gone. But not everyone is going begging. Amnesty International, Greenpeace and the United Negro College Fund announced record earnings this week, due mostly to large anonymous donations." Could only happen in the movies couldn't it … or could it? As the credits for the film roll, one discovers that the film was advised by one of the mathematicians who created these prime number codes, Leonard Adleman, the A of RSA.

But for the mathematician, the greatest entertainment of the primes is trying to discover how Nature chose these enigmatic numbers. Mathematics is about the search for patterns. Yet the primes, the building blocks of numbers, look like a sequence with no pattern to them at all – a fact that the script writers of the American series *Sex and the City* are clearly sensitive to. When the four girls at the heart of the series are out in New York's Central Park discussing their tastes in men, Miranda discovers that she always goes for the same sort of men, they always fit a pattern. But Samantha on the other had has a very eclectic taste in men: "I don't have a pattern". Carrie declares "in math randomness is considered a pattern". "And I'm what they call a prime number" Samantha responds. Samantha's taste in men is a great modern metaphor for the primes: you just don't know when the next one is going to come.

For 2000 years, ever since Euclid proved that there are infinitely many primes, mathematicians have tried to crack the Enigma of these mysterious numbers. An account of the pursuit of this great Holy Grail is given in my book The Music of the Primes, published in Italian by Rizzoli. A BBC documentary of the same name also dramatizes the search (see http://www.open2.net/musicoftheprimes/ and http://www.musicoftheprimes.com). It is a story with its own narrative drive to rival any of the films or books that have played with the primes as characters. But this is an unfinished tale: we still await the hero who will finally unlock the last chapter in the mathematician's story of the primes.

# Writing About Ramanujan*

Robert Kanigel**

A biography, to me, is meant for readers who may know relatively little about its subject. A biography of Picasso, for example, will be read by many who may not already know much about him – perhaps little beyond the most general about his life and work. These are readers, in other words, who are not experts, specialists, or scholars, but "general readers".

Now, it should be plain that this is not the only sort of biography you could write. You could choose to write a biography of a scientist aimed at other scientists, or even at scientists only in a particular field. This is a very different sort of thing to do; we have here, then, two different writing projects, with fundamentally different audiences, methods, preoccupations, goals. Neither is inherently better than the other. But they are fundamentally different. The author of either kind must be aware at every moment of what he or she is doing.

My biography of Ramanujan – *The Man Who Knew Infinity*, or *L'uomo che vide l'infinito* – is plainly of the first type, not aimed particularly at number theorists or other mathematicians, but at a broad general audience of readers who may, or may not, have taken calculus, or complex variables, at the university; who may, conceivably, have not even taken algebra in high school. Who may, in fact, be almost mathematically illiterate.

What do I, as a biographer of Ramanujan, owe my readers?

The answer might be batted back to me with another question: Isn't your prime responsibility to Ramanujan, to your subject, to the facts, to the truth, to the story of this great man? Shouldn't you be asking, *What do you owe Ramanujan?* True enough, but this debt to truth is owed by any biographer, any writer of nonfiction; it goes without saying. I am raising now a new question, one faced especially by writers who address their work to general readers, one particularly vexing given that, when it comes to mathematics, "general reader" covers such broad ground.

What do I owe my readers? Because mathematics is so difficult for so many, it might be argued that all I owe readers of a mathematician's biography is the story of the mathematician himself. A biography is sometimes called " a life"; the American

---

* Adapted from a talk given at "Matematica e Cultura 2005," Venice, 19 March 2005.
** The author is the copyright owner of this contribution.

and British editions of my book are subtitled, *A Life of the Genius Ramanujan*. So why not, we might ask, simply agree that Ramanujan led a life so captivating that we needn't venture into his mathematics at all. Perhaps we need only leave the reader peering over Ramanujan's shoulder at an incomprehensible scribble of lambdas and epsilons, exponents and integral signs, assert that Ramanujan is a genius, and let it go at that, proceeding without further ado to his life.

In bare bones form, Ramanujan's life goes something like this: He was born in 1884. He grew up in Kumbakonam, in South India. He went to school. Through a mathematics tutorial text called *A Synopsis of Elementary Results in Pure and Applied Mathematics*, by George Shoobridge Carr, he found himself seduced by higher mathematics. He lost interest in all else. He failed out of school. But he kept at mathematics, began keeping a notebook. He found patrons who gave him a sinecure with the Madras Port Trust. He wrote to eminent mathematicians in England with pages of theorems, but was ignored by at least two of them. Finally, he was embraced by a third, G. H. Hardy. Hardy and another noted mathematician at Trinity College, Cambridge, J.E. Littlewood, went over the theorems-stuffed letter they got from him and decided it was the work of a mathematical genius. Over the initial objections of his family, Ramanujan traveled to England a little before the outbreak of World War I. For five years he collaborated with Hardy. Together they made remarkable contributions to number theory and other mathematical fields. Ramanujan was made a Fellow of the Royal Society. He came down sick. At the end of the war, he returned to India, where he died at the age of 32, acclaimed as the greatest mathematician India had produced in a thousand years.

Behind this bare outline, of course, lies the immense detail that accumulates around anyone's life and that makes writing a biography such a substantial research effort. In my case, this meant reading most everything about Ramanujan that was already in print in English, talking to mathematicians, visiting the sites in England and India that played a role in Ramanujan's life as well as that of Hardy, learning as much as I could about the India and England of that era.

**Fig. 1.** Ramanujan (1887–1920)

In telling Ramanujan's story, I certainly had rich materials with which to work. I had the India of a distant time and place, when communication between West and East was more problematic than it is today. I had the cold shadows of World War I that darkened Ramanujan's life in England. I had the intriguing question of Ramanujan's spirituality; was he a devout Hindu, who worshipped his personal goddess, Namagiri; or was all that a matter of indifference to him, his spiritual observance, as Hardy insisted, no more than mechanical? I had, too, the endlessly fascinating relationship between the self-taught genius from India and the idiosyncratic Hardy. And, of course, the stark tragedy of Ramanujan's early death.

There was, in other words, no shortage of flavorful materials with which to tell of a life that was, even without the mathematics, rich, eventful, and full. If we ask, what do I owe my readers, perhaps the answer is, *All of this is quite enough, thank you.*

But would it be enough to write a biography of Picasso with barely a mention of Cubism, without referring to *Guernica,* without some insight into the way he painted, and why, and how? Would you want to read a biography of the architect Frank Gehry and learn nothing about his buildings or how he created them? Or a biography of Milan Kundera with scarcely a reference to his novels? The idea is absurd. How, conceivably, apply it to Ramanujan?

But here we come to the essential problem in writing about a mathematician. You can go to the museum and see Picasso's paintings, explore Gehry's buildings in Los Angeles or Bilbao, read Kundera and know what he's talking about. Mathematics, however, is different. Mathematics, we're told, is inaccessible. Mathematics, we learn from our friends in the humanities, is incomprehensible to all but math geniuses. Here is how I said it in the prologue of my book:

> Mathematics … is mired in a language of symbols foreign to most of us, explores regions of the infinitesimally small and the infinitely large that elude words, much less understanding. So specialized is mathematics today … that most mathematical papers appearing in most mathematics journals are indecipherable even to most mathematicians.

I went on to tell my readers that one of my informants, George Andrews of Pennsylvania State University in America, who rediscovered a long-forgotten Ramanujan manuscript in England, said it was only because he happened to be expert in one particular area of mathematics that he could recognize it for what it was. An ordinarily accomplished mathematician without that expertise would have been as befuddled as you or I. "What hope, then," I continued, "has the general reader faced with Ramanujan's work?

> Little, certainly, if we set as the task to follow one of Ramanujan's proof through twenty pages of hieroglyphics in a mathematics journal – *especially* in the case of Ramanujan, who routinely telescoped a dozen steps into two, leaving his reader to find the connections. But to come away with some flavor of his work, the paths by which he got there, its historical roots? These pose no insuperable problem – certainly no more than following a philosophical argument, or a challenging literary exegesis.

In one respect, writing about Ramanujan's math was a little easier than writing about some other areas. Because, I explained,

> much of it comes under the heading of number theory, which seeks out properties of, and patterns among, the ordinary numbers with which we deal every day; and 8s, 19s, and 376s are surely more familiar than quarks, quasars, and phosphocreatine. While the mathematical tools Ramanujan used were subtle and powerful, the problems to which he applied them were often surprisingly easy to formulate.

Or so I claimed, early in the book, on page 6, not wanting to lose more readers than necessary. But in practice? How could I actually write sentences, paragraphs, and pages that would give some flavor of Ramanujan's mathematics to readers scarcely schooled in mathematics at all?

One hint lies in that reference to "Ramanujan's mathematics." Some of the mathematics an average reader would need to understand *isn't* "Ramanujan's" mathematics at all, but ranges over intellectual terrain that is far simpler – the idea of an equation, of what is meant by proof, of basic mathematical manipulations. A biographer aiming his book at mathematicians might scoff at bringing up matters so elementary in a book about Ramanujan. But the writer for general readers must assume much less, must keep his readers, and what they do and don't know, ever in mind.

Early in the book, I referred to the influence on Ramanujan of Carr's *Synopsis*. And I suggested that the way it listed results without much in the way of proof encouraged Ramanujan to prove things on his own. I concluded this section by distinguishing the sometimes passive learning of many a high school student, led step by step through a proof, to the active stance Ramanujan brought to it: "Mathematics is not best learned passively; you don't sop it up like a romance novel. You've got to go out to it, aggressive and alert, like a chess master pursuing checkmate."

At one point, I described the mathematical housekeeping operation of "grouping like terms" as something like clustering mathematical entities in their appropriate categories. "You place dirty clothes in the laundry bin, freshly laundered napkins in the linen drawer, cereal back in the pantry. You put things where they belong." This is not remotely "Ramanujan's mathematics," but mathematically naive readers of the book will need a little of it. Not to understand Ramanujan's work. But simply to step, however uncertainly, into Ramanujan's world.

As the biographer of a mathematician, is my job to teach mathematics? I don't believe it is. And yet, if I can leave the reader feeling a little more at home in the potentially alien thicket into which I've drawn him – by explaining a little, pointing the way – I do owe him that. While I don't teach, perhaps my reader learns.

"Imagine," I asked my reader, "imagine cutting a hot dog into disc-like slices."

> You could wind up with ten sections half an inch thick, or a thousand paper-thin slices. But however thin you sliced it, you could, presumably reassemble the pieces back into a hot dog. Integral calculus, as this branch of mathematics if called, adopts the strategy of taking an infinite number of infinitesimally

thin slices and generating mathematician expressions for putting them back together again – for making them whole, or "integral."

I addressed this bit of mathematics because, in his letter to Hardy – the one that famously begins, "I beg to introduce myself ..." – Ramanujan included a number of definite integrals that he asked Hardy to consider. But on the strength of my account, could the reader perform even the simplest first-year calculus integration problem? No, that wasn't the point. Would the reader come away with a clear understanding of just what made Ramanujan's integrals so interesting to Hardy? I'm afraid not. Would he come away with some slight feeling for the *subject* of Ramanujan's thinking, the terrain over which his intellect roamed? Maybe, just a little.

Ramanujan's was a world of pure mathematics with little evident use in the everyday world. At least in the United States, but I suspect elsewhere as well, schoolteachers often feel the need to justify mathematics on the basis of its utility. *Learn algebra or geometry because it's useful* – useful not now, perhaps, but some day. And the drumbeat continues in most university courses. *Calculus gets us into space. Differential equations helps us design steam turbines.* But the number theory that consumed Ramanujan was "useless," almost proudly so; here was mathematics justified not for its utility, but for its own sake. Wrote G.H. Hardy, that foremost explicator of mathematics as an aesthetic principle. "No discovery of mine has made, or is likely to make, directly or indirectly, for good or ill, the least difference in the amenity of the world." The idea goes so counter to the suppositions held by non-mathematicians that it demanded to be addressed in my book, though in doing so, I offered no formulas, theorems, or proofs.

And "Ramanujan's mathematics" itself – that is, the mathematical topics with which his name will be forever associated? Like highly composite numbers? Infinite series? Partitions? Mock theta functions? Many readers with far greater mathematical understanding than I would have considerable difficulty with them. How do I know this? I know it because distinguished mathematicians told me so. For the most part, the understanding I gained of these and other subjects came only secondarily from reading the original papers. I needed help, and got it – from three prominent Ramanujan scholars, Dick Askey, Bruce Berndt, and George Andrews.

Another big help was Hardy himself. In his twelve lectures on Ramanujan he took his listeners into some of Ramanujan's work. But while highly technical, and aimed at other mathematicians, some of it was accessible to me. "Journalism," Hardy would write, "is the only profession outside academic life, in which I should have felt really confident of my chances." Hardy was a wonderful writer, about mathematics as much as anything else, a fact I made some fuss about in the book. "He wrote, in his own clear and unadorned fashion, some of the most perfect English of his time," C.P. Snow once said of him.

Still, with all the help in the world, I have my intellectual limitations, and my readers have theirs. Does this mean it's time, finally, for the biographer to throw up his hands in surrender, mention by name the topics of Ramanujan's study, as I just did, and be done with it? In fact, I tried to take my readers a little further than that. I tried to explain what a highly composite number is. What a partition is, and what

213

the problem that so attracted Ramanujan and Hardy was all about. And even to give the slightest flavor of one of their achievements, a powerful approximating method called the circle method that "let you draw oh-so-near, but never actually touch, the forbidden circular path," where a particular function was undefined.

Now, from many a professional mathematician's standpoint, these feeble offerings might be dismissed as hopelessly superficial, the mathematical bar set way too low. But just how high, or low, are we to set it? What is the "proper" level? The question, of course, has no correct answer; it's one you could debate all day and all night. While many mathematicians might come away frustrated with my account, one that fails to convey much of the detail and nuance of Ramanujan's work, an account more successful at that level would, in turn, lose practically all ordinary readers.

Too bad, the objection might go. That's what Ramanujan did, isn't it? Anything less is piffle and froth. It doesn't matter if the reader doesn't understand.

*Doesn't matter?!?* To the science writer, *that* represents the abysmally low standard; it's a kind of literary capitulation. Mathematicians do mathematics? Well, science writers – those who write professionally about science, technology, and medicine for ordinary readers – worry and fret and struggle with words on behalf of their readers. Writing about mathematics is, to state the obvious, a different discipline from doing mathematics. And while not so difficult in most respects, perhaps, yet it is more challenging, sometimes maddeningly so, in others.

Most science writers are, like me, generalists, routinely in a position of not knowing, not being expert, not themselves possessing the educational and professional tools that experts in biochemistry, or information theory, or molecular biology, or number theory do have. And yet, they set out every day to write the newspaper articles, magazine articles, and books you read. Overcoming armies of their own ignorance, they use their victories over it to help their readers understand and appreciate difficult material from which otherwise they'd be wholly excluded. Dare to write about the great Ramanujan without being mathematician enough yourself to do his work true justice? Well, this, or something like it, is what science writers do, with quarks and nucleotide bases and fractals, every day; it is an occupational hazard of the field.

Often, the science writer's work requires introducing the reader to material that specialists learned not in graduate school, not in undergraduate school, but which goes back so far they may have no memory of having ever learned it at all. For example, how do you take readers into the seemingly nonsensical territory of numbers like the square root of minus one – imaginary numbers? Much of Ramanujan's work with Hardy existed on the complex plane. Well, I turned to that first departure most of us have from the ordinary numbers we use every day, negative numbers. I introduced imaginary numbers, admitted that they represented something of an alien and unlikely idea, but then wrote:

> That happens often in mathematics; a notion at first glance arbitrary, or trivial, or paradoxical turns out to be mathematically profound, or even of practical value. After an innocent childhood of ordinary numbers like 1, 2, and 7, one's initial exposure to negative numbers, like $-1$ or $-11$, can be unsettling. Here, it

doesn't require much arm-twisting to accept the idea: If $t$ represents a temperature rise, but the temperature *drops* 6 degrees, you certainly couldn't assign the same $t = 6$ that you would for an equivalent temperature rise; some other number, $-6$, seems demanded. Somewhat analogously, imaginary numbers – as well as many other seemingly arbitrary or downright bizarre mathematical concepts – turn out to make solid sense.

Of course, this is all embarrassingly simple to any working mathematician. But it's not working mathematicians to whom I directed *The Man Who Knew Infinity*. And while some mathematicians might need help understanding the difference between a quasar and a quark, a tannin and a bate, the double helix of DNA and the triple helix of collagen, many ordinary readers would need similar help with imaginary numbers. The biographer of a mathematician has the job, as does any science writer, of introducing his readers to what to them may seem an alien, intellectually daunting world.

But perhaps I shouldn't use the word "introduce," with its connotations, at least in English, of fulfilling a social obligation at a party and then, perhaps, wandering off to another part of the room. Because in writing for ordinary readers it's important to remain their steady companions, to stand at their side for the whole evening, for the length of the book.

This, at any rate, is what I tried to do in my biography of Ramanujan. In doing so, in bringing his life, his work, his world, a little closer to the rest of us, I hope that, in a small way, I honored his memory.

215

# Maat and Thalia*

Maria Rosa Menzio

MAAT is universal order according to the inhabitants of ancient Egypt (not the god of mathematics). THALIA is the Muse of Comedy (there is no Muse of Drama). Now, what to mathematics and drama have in common? My answer is: emotion.

In fact, writer and mathematician Denis Guedj asks:

> Where is the emotion in science? How is it manifest? What testifies to it? In short, what relationships are there between truth and emotion?

> History of science is full … of stories of science in which truth feeds fiction, rigour subtends narration. The sciences profoundly shape modern society [1].

But only in the last few years has science entered into theatrical scene. Until a short time ago it was, in fact, only rarely an instrument of drama (one rare example was Brecht's "Galileo").

> But since the most ancient times, society has had story-tellers. They perform a social and individual function; they appeal to the imagination, but also to tastes. The field of knowledge, especially that of scientific knowledge, can be a formidable field for drama.

The historian of science has before him research and documents that sometimes leave gaps between two known events. As a scholar, he cannot invent. The facts are the strong points of his work, but they are also his limits. However, there is someone who can go beyond those limits: the scriptwriter. He can fill the gaps, imagine facts that are not true, but that have the ring of truth to them. If there is one historic fact A and another B, both documented, but without the "truth segment" AB, then the dramatist can invent the event AB that is fictional but possible.

> He creates "between the lines" and does what creates the desire to read what he writes: he invents a universe. [...] Real fiction. Fiction, because it is the au-

---

* *Translated by Kim Williams*

thor's imagination that determines its value; real, in as much as it conforms to scientific and historic truth.

For non-specialists, mathematics has always had the fascination of absolute truth. But some questions arise here. Is it possible to speak about a mathematical theorem as we do about any other historic event? And the heroes of mathematics, the greats? How can we speak of them if their research is not yet finished? How much freedom do these heroes have? How much "free will" do they have on the stage?

## Time

From "IL MULINO" (The Mill) (National premiere 20 May 2005)

Her: Things are numbers. So says Pythagoras. And mathematics is born. The planets are gods. Astronomy is born.

Him: The poet searches, searches, dips into the immense cauldron of myths. He is the prospector of fables. He sees rough figures, lively, the Scandinavian peoples. There is Amlode, the hero of the legend, Amlode the sad, who lives in a fabulous mill …

Her: *The sky is all green, and the fields are silver.*

Him: In those times, the mill was different, it ground peace and abundance.

Her: *The clouds are indigo, and the meadows are salmon coloured.*

Him: Salt … Then, decadence, and the mill grinds salt. Salt …

Her: *The heavenly vault is violet, and the heavenly crops are gold.*

Him: Death … death … The mill has fallen,
Now it is at the bottom of the sea, and grinds rocks and sand.
There is a whirlpool, a giant whirlpool, whose name is *maelstrom*.
The Umbilicus of the Sea. The way that leads to the Realm of the Dead.

Her: Whose fault is it? Of a giantess, thrice born, thrice killed, but living still. One night she stopped the mill, she sung a terrible song.

Him: Look up! At the Northern borders there are the Seven Sages, the Seven Stars of Ursa, Lady of the Revolving Heavens.
The Seven Oxen that go round the millstone of the mill.
The Galaxy is a Bridge, that leads out of Time.

Her: The Master of the Dance performs new steps and creates Ursa Major.
Mathematics and astronomy establish the myth.
Hark! The Time of Music is coming.
It has the pace of a king, and walks between Cycles and Epicycles
It comes from the Prime Mover, the Immobile Mover.
Not even … not even a hair between the calendars and the notes.

### Seven Musical Notes

Measures of time and music of rites: agreement, perfection
The blows striking the anvil make the music.

## Music

The solar year and the octave dominate the world. Number and Time … the flow of Time …
The time that runs with seven reins and a thousand eyes
　　And seven wheels, and the axle is Immortality.

Him:　Gods and Humans, Trees and Animals, Crystals and wandering Stars, all...
　　　Only two masters: Law and Measure
　　　The seal of Time, the First and Last Seal, the alpha and omega.
　　　*The world was all a vengeance, a boiling sea, the land had no space to walk in, the air was assaulted by lances …*

## Battle noises

PEACE … SALT … DEATH
*The stars began to battle! Earth and time washed their hands in disaccord.* Peace … salt … death …

Her:　*The whole world is my realm, from Pisces to the head of Taurus.*
　　　From Pisces down to Aldebaran
　　　Thirty degrees of the zodiac, the stars of Aries.
　　　*His realm is not Heaven, but Time, the dimension of Heaven.*
Him:　As of today there will be new feasts and new customs
　　　For the King of Aries. Let the Ocean of Energies run!
　　　The course of the stars, sovereign, the symmetry, *fearful symmetry* of tigers and theorems … **Tigers and theorems! Tigers and theorems!**

219

From "SENZA FINE" (Without end) (National premiere 18 April 2004)

Hypatia:　Find a segment that is the side of a square equivalent to the circle!
Voice:　Time flows. It feeds on memories. It crouches between the temples and strolls with its soft steps. She was a famous mathematician, a ton of years ago, in Egypt, our Hypatia. And she had solved many a problem! But not that one. That one never came to her. Squaring the circle: "Find a segment that is the side of a square equal to the circle itself …!" And as if in a dream, she sees spirals and strange curves … spirals like those in a snail, or in a screwdriver, in a screw-driver, that over time grabs onto itself, it makes your head spin …
Hypatia:　Not I, I have no ankle bracelets, bracelets or King Solomon's rings – I have no golden haloes, nor power, nor glory – I am not a lost bride, a devastated city – I have no dragons nor poisoned apples – no enchanted forests – no magic garters – no circles of fire – I have no double rings that make infinity when put together, tomorrow I will have only – as a ring, a collar – the noose – of the executioner – and on the tree of the world – I shall be hung – hung – hung – hung – hung.
Voice:　"Travelling with a spaceship it is possible to visit any part of the past, present, and future!" That crazy woman, she exalts … flies … that Hypatia. A round flight. A flight that flows into a square time. She lands

in a meadow. Square. Special. A meadow … of four-leafed clovers. An emerald-green handkerchief of land. A symphony of gems hidden in a fold of the arrow of time.

Hypatia: "Lady, are you hurt? … lord…I am dead!"

Voice: There is a man, in the meadow where she trips, where she falls, on all fours, on her face, she gropes … moves … in time … through the days … through the months … through the years, flown by, again. But is it a man or a woman, in the time of six or seven lifetimes? There is Orlando, who is a man and then woman and who escapes from the circle. Who bends the doors to her geometry.

Hypatia: And the squaring of the circle?...The duplication of the cube?...The trisection of the angle? I turn the ring and think of tomorrow. Tomorrow is another day.

Voice: That's right. Gregorio XIII reformed the calendar. From the Julian to solar. To pass … directly from Thursday 4 October to Friday 15 October. Still one-thousand-five-hundred-eighty-two. Ten ghost days. Ten days … lost!

Hypatia: The cards! The Tarot! Look carefully into my eyes! … Choose … Temperance! The sun and the moon, on the serpent that devours itself. Homer's golden chain … and … Plato's ring. The globe at the centre of the earth. The Stone. The stone and the ring … sometimes it goes to my head … the stone of the Philosophers. The Ring of time! Alchemy! Magic!

Voice: What marvellous work! Inventing the astrolabe! The astrolabe!

Hypatia: Calculating the height of the stars with respect to the horizon. The stars, up above … and then to lose oneself somewhere there, outside of orbits and time … in the night …

Voice: You can't go beyond. There is another dimension, beyond.
The course is finished. You are annihilated. Hypatia … is in Hades.
And while the earth continues its course through space … Hypatia … is obliterated by time. Near the meadow of four-leafed clovers, the trunk of the magic apple tree, (the one that had all the colours of the rainbow), has become grey. Like lead. And I … I … am now like stone. Stone … stone … stone …

From "HENRY IV" by Luigi Pirandello [3]

Henry IV: You feel yourselves alive in the history of the eleventh century, here at the court of your emperor, Henry IV! […] And to think that at a distance of eight centuries from this remote age of ours, so coloured and so sepulchral, the men of the twentieth century are torturing themselves in ceaseless anxiety to know how their fates and fortunes will work out! Whereas you are already in history with me. […] And sad as is my lot, hideous as some of the events are, bitter the struggles and troublous the time – still all history! All history that cannot change, understand? All fixed for ever! And you could have admired at your ease how every effect followed obediently its cause with perfect logic, how every event took place precisely and coherently in each minute particular! The pleasure, the pleasure of history, in fact, which is so great, was yours […] The solitude […] I determined to deck it out with all the colours and splendors of that far off day of carnival, when you Marchioness, triumphed. So I would oblige all those who were around me to follow, by God, at my orders that famous pageant which had been – for you and not for me – the jest of a day! I would make it become forever – no more a joke; but a reality, the reality of a real madness: here, all in masquerade, with throne room, and these my four secret counsellors: secret and, of course, traitors. […] I am cured, gentlemen: because I can act the mad man to perfection, here; and I do it very quietly, I'm only sorry for you that have to live your madness so agitatedly, without knowing it or seeing it.

This is my life! Quite a different thing from your life! Your life, the life in which you have grown old - I have not lived that life!

From "TO EACH HIS MINOTAUR" [4] by Marguerite Yourcenar

Ariadne (*addressing herself to God*): You have centuries at your disposition, your time is measured in epochs that are practically eternal. But Theseus has at most fifty years ahead of him.

From "WAITING FOR GODOT" [5] by Samuel Beckett:

Pozzo: Have you not done tormenting me with your accursed time! It's abominable! When! When! One day, is that not enough for you, one day he went dumb, one day I went blind, one day we'll go deaf, one day we were born, one day we shall die, the same day, the same second, is that not enough for you?

And finally, from "MACBETH" [6] by Shakespeare:

To-morrow, and to-morrow, and to-morrow,
Creeps in this petty pace from day to day
To the last syllable of recorded time,
And all our yesterdays have lighted fools
The way to dusty death. Out, out, brief candle!

## The Numbers

From "FIBONACCI" (La Ricerca) [7] (National premiere 10 June 2003):

| | |
|---|---|
| Zaffira: | For hundreds of moons my family has had magic powers. |
| Fibonacci: | What powers? |
| Zaffira: | It falls to the eldest daughter, that is, to me, to weave a tapestry: the tapestry of past and future. |
| Fibonacci: | What do you mean? |

| | |
|---|---|
| Zaffira: | Each thread is a human life. The fabric that grows is time that flows. But the future is always lying in wait. |
| Fibonacci: | Is there a thread for yourself as well? |
| Zaffira: | You see these two red threads? They are our destinies. Our lives are interwoven, then they separate. Because of a tragedy. |
| Fibonacci: | This tapestry is cursed! It influences the future, makes happen what you want to happen. Your fabric frightens me! [...] |
| Him: | Is it the golden section? |
| Her: | So. You take two Fibonacci numbers one after the other. |
| Him: | Like eight and thirteen. |
| Her: | Or thirty-four and fifty-five. And divide the first by the second. Eight divided by thirteen, or thirteen divided by twenty-one. The result is $x$. Well then, while the Fibonacci numbers become ever larger, this "$x$" result ... |
| Him: | The result of the division. |
| Her: | That's right, this number "$x$" becomes ever closer to the number 0.618 ... that is then (*triumphant*) the golden section of a segment. The world cannot have been made randomly. |
| Him: | No, wait. You go too fast. If our segment is called "a", the golden section will be our unknown "$x$", (*writing on the blackboard*) and so we have the proportion $a : x = x : (a - x)$. |
| Her: | $a : x = x : (a - x)$. Exactly. And the golden section of a segment is more or less equal to two-thirds of the segment itself. |
| Him: | It's an aesthetic fact. If you have a nice dress and you want to put a ribbon on the belt, where do you put it? At two-thirds of the belt, because it looks good there. (*He tries to grab her by the waist*) |
| Her: | Don't touch me! |
| Him: | I'm only trying to make you understand. The ribbon goes right at the golden section. |
| Her: | Watch your hands. |
| Him: | Oh brother! [...] |
| Zaffira: | It can only happen at the winter solstice. The gate of the gods. It is an ancient secret of magic, but you have to want it with all your might. |
| Fibonacci: | Want what? |
| Zaffira: | You know, that day time rolls onto itself. All that has been becomes a dream. As if it never happened. |
| Fibonacci: | What do you mean, that you can cancel a piece of a life? |
| Zaffira: | I mean that at the winter solstice, on the altar of sacrifices, there is a knot of the arrow of time. Centuries pass in an instant ... |
| Fibonacci: | But what is this altar? |
| Zaffira: | It is that stone in the middle of an oasis. The old ones say that in the past it was used for human sacrifices, when the gods were miserly with water and asked for blood. |
| Fibonacci: | And don't they ask for blood anymore? |
| Zaffira: | Now the gods don't care. We have to make it on our own. [...] |

| Him: | I wanted to ask you to have dinner with me tonight ... |
|------|--------------------------------------------------------|
| Her: | Only to have dinner? |
| Him: | Get out, you're so suspicious! I made chicken ala Fibonacci |
| Her: | Chicken ala Fibonacci? Ahahah! And how is that? |
| Him: | With a litre of beer. |
| Her: | Mmmmmmmh. |
| Him: | 618 grams of onion. |
| Her: | No, no onion. |
| Him: | Yes, you have to have onion. |
| Her: | Mmmmh. And then? |
| Him: | A lemon. And then 618 black peppercorns. |
| Her: | And you count them all yourself? |

## Possible Worlds

From "A FINE PARISIAN DRAMA" [8] by Alphonse Allais (1855–1905)

The protagonists of the story are Raoul and Marguerite, two newlyweds. Their life together could be considered a happy one, but they both had a temper ... In short, each of them always wanted to be right. Plates were thrown, and fists flew.
One night our heroes went to the theatre, where they saw the play *L'Infedele (The Unfaithful)*.
One morning Raoul received the following message:

"A word to the wise. If you wish to see your wife happy for once, be at the Incoherents' Ball next Thursday at the Moulin Rouge. She will be there, masked and disguised as a Congolese Pirogue ... A Friend."

That same morning, Marguerite received the following message:
"A word to the wise. If you wish to see your husband happy for once, be at the Incoherents' Ball next Thursday at the Moulin Rouge. He will be there, masked and disguised as a turn-of-the-century Templar ... A Friend."
These words did not fall on four deaf ears.
Admirably cloaking their designs, when the fateful day arrived:
"Dearest," said Raoul in his most guileless tone, "I shall be forced to leave you until tomorrow. Affairs of the utmost importance require my presence in Dunkerque."
"It's just as well," replied Marguerite, delightfully candid. "I've just received a telegram from my Aunt Aspasie, who is quite ill and has called me to her bedside."

*(The reader thinks of world A in which the two betray each other)*

The society columns of the *Limping Devil* were unanimous in proclaiming that this year's Incoherents' Ball shown with unusual splendour.
Many bare shoulders and a fair number of legs, without mentioning their accessories. Two of those present did not seem to be taking part in the general merriment: a turn-of-the-century Templar and a Congolese Pirogue, both hermetically masked.

At the stroke of three in the morning, the Templar approached the Pirogue and invited her to join him for supper. In response, the Pirogue merely lay her delicate hand on the Templar's robust arm, and the couple departed.

"We want to be alone," said the Templar to the waiter of the restaurant. "We will choose what we want to eat and then we will call you."

The waiter left and the Templar locked the door to the room. Then, with a sudden movement, after having rid himself of his helmet, he ripped off the Pirogue's mask. Both of them, at the same time, emitted cries of stupefaction, as neither recognized the other.

*He* was not Raoul.

*She* was not Marguerite.

They offered each other their apologies, then lost no time in making each other's acquaintance over a light supper, and that's all I have to say to that.

*(The reader discovers that the two masqueraders don't know each other, that it is not he and she, and so believes that the author is speaking of another world, B)*

The small mishap served as a lesson to Raoul and Marguerite [...] They never argued again, and lived happily ever after. They don't have any children yet, but they will, you'll see, they will.

*(The author shuffles the cards and says "this serves as an example", thus returning to world A. And the reader is justifiably disoriented)*

In *Lector in fibula* Umberto Eco suggests that the reader has produced impossible worlds through his own expectations, and has discovered that these worlds are inaccessible to the world of the narrative. But the narrative, after having judged these worlds to be inaccessible, returns to them. How? Not by reconstructing a world with contradictory properties, but thinking that these inaccessible worlds could be in contact. I would add that perhaps the author amused himself by constructing a story on the expectations of the reader, on his world full of logic. The reader always imagines himself to be dealing with an invented world, but not a contradictory one, or one with riddles that can't be solved. So then, Alphonse Allais breaks with this tradition and is ironic about the reader's certainties. He stands on the sidelines in order to watch, smiling, the spectator's reaction. He hears him say, "It's impossible!" Or, "How is this going to end?"

## Equations and Triangles

From "INCHIESTA ASSURDA SU CARDANO" (An absurd enquiry into Cardano) (National premiere 16 September 2005)

Bianca:   We held hands as we rode on horseback. Gerò and I. We could break our necks. Then ... the spell is broken, the fatal cannonade comes, and I fall, the distance separating me from the earth is long ... long and short. I don't want to be to trampled, I scream. Because there is so much light. But now

I can't see anymore. I want this night to end. Evil comes like a serpent. Down, from the throat down the back. Falling headlong down the stairs. Plinths. I have fallen. I will never play with him again. No more dances, no more love. I was a virgin then, and whole. After, the fall. Lame, and a courtesan.

Gerò: (reading) Now then, let's see.
When the cube and things together
Are equal to some discreet number,
Find two other numbers differing in this one.
Then you will keep this as a habit
That their product should always be equal
Exactly to the cube of a third of the things.
The remainder then as a general rule
Of their cube roots subtracted
Will be equal to your principal thing [9].

Bianca: Stop already with your predictions!

Gerò: I made the horoscope myself. I have a large sheet, gridded, a metre long and a metre and a half wide, and inside there are seventy-five little squares, five by fifteen, and I will die at seventy-five years old, no more, no less.

Bianca: Stop it!

Gerò: I was born in September and I will die in September. I'll go mad!

Bianca: I'm sure you will! You are always looking at that damned sheet, and think that that is your seventy-fifth year, already passed, irrevocable …

225

Gerò: The Etruscans lived like that too. There was a secret wall, where they hammered in nailed, and every nail was a year and there were just that many nails and not a one more, and they looked at the space that remained! It was always smaller!

Bianca: Gerò! I love you and you don't even care!

Gerò: Destiny. The black squares, the squares passed, lived, lost!
[…]

Bianca: I had known it already for some time, that you wanted him and not me.

Gerò: And you didn't say anything to me. Is there anything else I don't know?

Bianca: You, me, and my brother, what a nice triangle … truly cabalistic!

Gerò: What the devil are you saying?

Bianca: Him, in love with me, what a shame, to want your sister. I in love with you, what a scandal, a married man! And you enamoured of him, how indecent! Homosexuality! A trio of sinners, going straight to hell!

Gerò: But nothing has happened! Never anything!

Bianca: Indeed! It went wrong for everyone. I never went to bed with you, you never went to bed with him, and he never went to bed with me. Third-rate sinners.

Gerò: And now it's too late.

Bianca: Yes, that's right. Too late to commit it, that sin that we always dreamed of, each his own. Passing through life dreaming of the impossible. A trio that closes.

Gerò:     Nice triangle. And nothing ever happens.
          [...]
Bianca:   Only after seventeen generations will the truth be known. The truth, that
          Morgan le Fay, the truth, about me, about Cardano, the truth, that chimera
          ... the truth ... that faraway mirage *(Complete darkness)*

## Bibliography

[1]  D. Guedj (2001) *Il Meridiano*, Longanesi, Milan.
[2]  M.R. Menzio (2005) Senza fine, in: M.R. Menzio, *Spazio, tempo, numeri e stelle*, Bollati Boringhieri, Turin.
[3]  L. Pirandello (1922) *Henry IV*, Trans. Edward Storer, E.P. Dutton, New York.
[4]  M. Yourcenar (1984) To Each his Minotaur, Trans. Dori Katz, in: *Plays*, Performing Arts Journal Publications.
[5]  S. Beckett (1954) *Waiting for Godot*, Grove Press, New York.
[6]  W. Shakespeare, *Macbeth*.
[7]  M.R. Menzio (2005) Fibonacci (la ricerca), in: M.R. Menzio, *Spazio, tempo, numeri e stelle*, Bollati Boringhieri, Turin.
[8]  A. Allais (1997) A Fine Parisian Drama, in: *Anthology of Black Humor*, ed. Andre Breton, City Lights Books, San Francisco.
[9]  Tartaglia's poem, Eng. trans. J J O'Connor and E F Robertson. http://www-history.mcs.st-andrews.ac.uk/HistTopics/Tartaglia_v_Cardan.html.

# Mathematics and Cinema

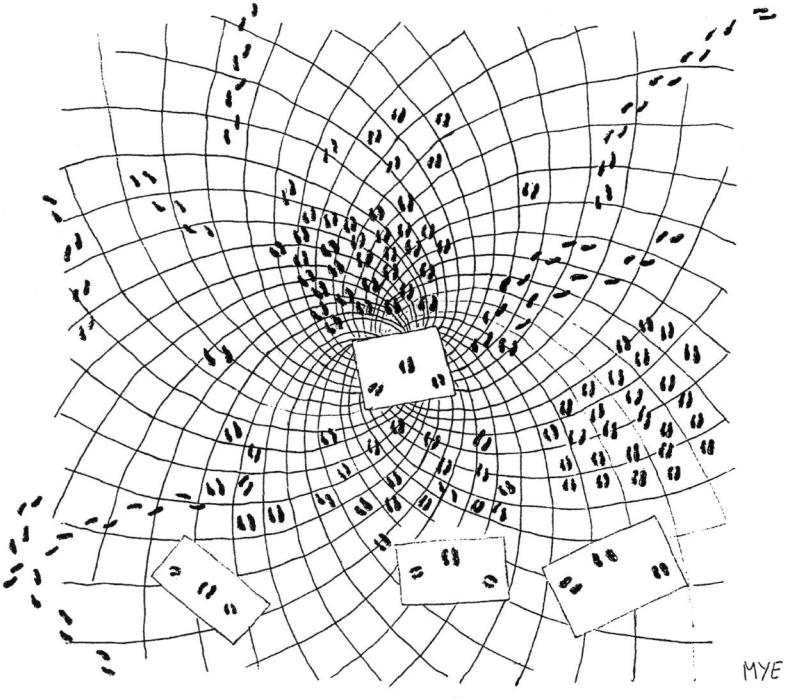

MYE

# *Assioma 5*: A Scientific, Mystical, Historical Film*

Adolfo Zilli and Elisa Cargnel

In September 2001 a group of friends started to think about the idea of making a short film that would address in a scientific way one of the fundamental tenets of Oriental religions: the omnipresence in the universe of a physical quantity called energy.

It was inevitable that Einstein should be evoked, since, with his proof of the equivalence of matter and energy, he had given a new interpretation of space and time. In order to achieve that, Einstein had to eliminate one of the pillars of geometry: Euclid's fifth axiom, accepted as true but not proven, which states that two straight parallel lines never intersect.

Some mathematicians (Riemann and Lobačevskij in particular) had already studied this hypothesis for purely speculative and theoretical purposes. Some artists had offered more or less knowing interpretations of these concepts: many of Escher's graphic works and Bach's *Musikalisches Opfer* are some examples of artistic applications of non-Euclidean geometries.

Hofstadter, in the book *Gödel Escher Bach: The Eternal Golden Braid* [1], carefully examines these experiments, weaving them into a single artistic thread and defining them as "crabs", a term which indicates the interweaving of a single theme with itself, simultaneously developing forward and backward.

The idea of using cinematographic fiction to relate the equivalence of matter and energy, together with the desire to make a "crab" film resulted in *Assioma 5 – oppositi e paralleli* (*The Fifth Axiom – opposites and parallels*), a medium-length film that tells, romanticizing them, the dual stories of Einstein and Euclid. Events that interweave, interpenetrate, and end in the formulation of the fifth axiom for Euclid and in the proofs and theory of general relativity for Einstein.

The present paper will begin by explaining in greater detail the relationship between relativity and the fifth axiom, and then will go on to show how the film brings out a constant duality between Einstein's character and that of Euclid, even when both decide to change their own characters. Finally, we will see how these ideas are

---

* *Translated by Kim Williams*

inserted into the film according to a "crab" structure, and at the same time look in depth at the meaning of this definition.

The final section will discuss how our work group was organised and the method with which we achieved the unanimity required for every decision.

## Relativity and the Fifth Axiom

The theory of relativity was formulated at two separate moments and is composed of two theories that complete each other.

The special theory of relativity was dealt with in an article that Einstein published in 1905, where he had already formulated the equivalence between mass and energy with the famous formula $E = mc^2$. As innovative and plausible as it was, experimental proof of the theory had to wait until 1919, when an expedition to Prince Island, led by the English astronomer Eddington, furnished the photographic plates of a total eclipse that proved the truth of the theory of relativity (Fig. 1).

These plates make it possible to see a star, which according to astronomical calculations should have been hidden by the disc of the sun.

The step forward taken by Einstein in the second part of his work was to think of space-time as a cloth curved by a mass resting on it. In this case a ray of light, which in theory moves in a straight line, is influenced by the gravitational fields of the heavenly bodies in the same way as a marble shot across the cloth would be influenced by the mass resting on the canvas itself.

This confirmed the hypothesis that light, even though considered to be pure energy, behaves as a mass in that it can be influenced by gravity; but this was not the only result. As a consequence of these ideas, when giving a mathematical expression for a physical law it is necessary to keep in mind that the straight line is no longer a rigid, invariant entity, lying on a flat plane, because the very nature of space itself where it is found has been modified. For this reason, many mathematical laws are no longer valid, since they derive, directly or indirectly, from the assumption that space is rigid and cannot be curved.

**Fig. 1.** Frame from the film, Expedition to Prince Island

In order to find out which laws are still valid and which can be eliminated, it is necessary to rethink everything, starting from the basis of geometry, and more precisely, from the Euclid's fifth axiom, which does not admit that two parallel straight lines can have a point of intersection. Already a century before the theory of relativity, Lobačevskij had defined a non-Euclidean geometry that denied the parallel postulate and defined a segment inscribed in a circle a "line", and the circle itself as a "plane". Other kinds of non-Euclidean geometry were defined successively by Gauss and Riemann, and it was by following the model of hyperbolic geometry proposed by Riemann that Einstein arrived at the proof that our universe follows precisely the laws of a non-Euclidean geometry, thereby disproving a theory that had been undisputed in Western culture for more than two millennia.

## Einstein and Euclid, Two Historical Figures, Two Parts of the Human Brain

Euclid lived in the third century B.C. in Alexandria, Egypt, during the Hellenistic period. Only a handful of writings exist as testimony to him, if we leave aside the thirteen books of the *Elements* that he, with his school, left to us. This treatise is a compendium of all the geometrical knowledge collected in Alexandria, without a doubt the city that was the greatest cultural centre of the entire ancient world.

The laws of geometry assembled by earlier cultures were often based, not on numbers, but on practical rules for navigation, construction of houses, and the division of lands and harvests. The unification of these rules required a higher degree of abstraction, so that fields and walls were replaced with lines, triangles, and circles. The theory that Euclid articulated had to analyse the common aspects of these rules and deduce general laws from them, while constantly making sure that no single statement contradicted what had been done previously.

Even if the *Elements* a clear example of analytic and deductive rational thought, as certainly befits one of the Aristotelian school, it can't be said that this mentality was very widespread, except within the academic world. The great number of myths

**Fig. 2.** Frame from the film, the eclipse in ancient Greece

231

and legends that appeared to govern the world of that time were often contradictory or irrational. Think, for instance, of how ordinary people at the time might have interpreted events that are easily explainable for us, such as a total eclipse of the sun (Fig. 2).

Albert Einstein, in contrast, grew up with a special interest for whatever turned out not to be easily explainable for science, and this led him to undertake scientific studies, even if his scores were not always the best, probably due most of all to his rebellious nature, which hindered him from accepting the rigour and the arrogance of those milieus.

At the Zurich Polytechnic he met his future wife, Milena Mariè, also a very good physicist, who in any case represented an ongoing connection with scientific knowledge after Einstein abandoned the world of the university. On the other hand, shortly before formulating his theories, Einstein had formed a reading group with some friends (*Akademia Olympia*), which was an outlet for every kind of doubt about the bases of rational thought, mathematics, physics, giving preference instead to arguments that were humanistic or narrative.

There is an obvious difference of mentality between the two figures (Fig. 3), but also a complementariness that can be likened to the diversity between the two hemispheres of the human brain: the right hemisphere is the seat of the emotions, creativity, and imagination, while the left hemisphere is the seat of rationality and logic. Defining this concept with two terms that belong to Oriental philosophies, the film *Assioma 5* underlines the fact that Einstein had a predominantly *yin* nature (emotional), while the mind of Euclid was of a predominantly *yang* nature (rational), even if within each mentality we find a component of the other one. This concept is aptly represented by the *Tao*, the symbol that indicates duality in the Orient (Fig. 4).

Note that both figures end up completing their own nature by embracing the opposite: for Einstein it becomes important to describe mathematically and prove scientifically his purely philosophical thinking; in contrast, for Euclid it become important to make statements without having to give their proof.

**Fig. 3.** Albert Einstein and Euclid of Alexandria

**Fig. 4.** Tao, the symbol of dualism in Oriental philosophy

**Fig. 5.** Frame from the film, Milena Mariè, Einstein's wife

**Fig. 6.** Frame from the film, Apollonius, Euclid's student

In writing the script for the film, in order to allow the two characters to change their opinions, it was necessary to flank them with secondary characters who provided an opposite vision, as if Einstein's alter ego talked to Euclid and vice versa. Thus Einstein's wife (Milena Mariè, Fig. 5) pushes him towards academic proofs, and a fatalistic student of Euclid's (Apollonius, Fig. 6) opens the horizons of axioms for the master.

To the two stories we added an arrow that passes through time and space; this arrow is shot by an archer in ancient Greece at the end of the first act, when the conflict between both characters is defined, and continues to travel throughout the whole second act, finally making a bull's-eye in Einstein's pendulum clock in his dream at the beginning of the third act, thus representing for both the resolution of the conflict.

## Crabs

In his book, Hofstadter lists a conspicuous number of arts based on non-Euclidean geometries, beginning, as the title suggests, with Bach, Escher and Gödel, but going on to a fascinating examination of other examples of similar applications.

In order to clarify the structure of the film, we begin with Bach's *Offerta Musicale*, which is probably the most intuitive example and the one that brings us the closest to what we applied in constructing the script. Bach begins with the musical theme that was proposed to him; he tries playing the notes from last to first, thus finding himself with two sequences of notes. The superimposition of these two sequences, with rhythmic adjustments that serve to give to both a harmonic significance, constitutes an example of a "crab" score. In actual fact, Bach's score is much more complex: onto the two themes, that proposed to him and the same one played backwards, are superimposed four more copies (canons) of the theme and another four played backwards (inverse canons), each displaced by a given interval of time with respect to the first.

Another example of "crabs" can easily be discerned in the xylographs of Escher (Figs. 7–8). In these works there are two levels that are superimposed, and one level moves in a direction opposite that of the other.

To build the structure of the film, we started from one of the classic scriptwriting schemes, Syd Field's paradigm, which calls for a division of the story to be told in three acts and two pivotal points:

– the first act involves laying out the story and the presentation of the characters;
– the passage from the first to the second part is defined by an incident's being set off, one that will create conflict in the story;

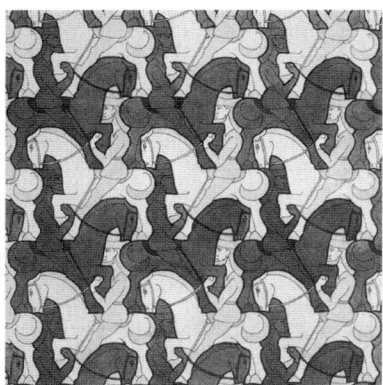

**Fig. 7.** M.C. Escher, Symmetry Drawing E67 (*Horseman*). © 2006 The M.C. Escher Company – Holland. All rights reserved, *www.mcescher.com* (*see the section in colour*)

**Fig. 8.** M.C. Escher, Symmetry Drawing E117 (*Crab Canon*). © 2006 The M.C. Escher Company – Holland. All rights reserved, *www.mcescher.com* (*see the section in colour*)

- in the second act the obstacles that must be overcome in order to resolve the conflict are faced;
- the passage from the second to the third part is defined by an event that resolves the conflict;
- in the third act a new equilibrium is created after the resolution of the conflict.

Even while proceeding in parallel with this scheme, the stories of Einstein and Euclid have been studied so that each event that happens to one character happens to the other one as well, but in an opposite position with respect to the central moment of the film, which for both characters takes place during the second act.

235

The conflict, for Einstein, consists in the scientific proof of a "vision" that had convinced him of the equivalence of matter and energy; the conflict for Euclid, on the other hand, is that which gives a motive to a thesis that, even if not proven, lies at the basis of his work: the parallel postulate.

Not much is known of Euclid's life; this has allowed us to choose the most interesting events from Albert Einstein's life and to construct on the basis of these the story of Euclid, transporting them to ancient Greece. Thus, for example, a party among friends becomes a symposium to celebrate the final draft of Euclid's *Elements*; in an analogous way, the eclipse that proved Einstein's theories becomes an eclipse that Euclid is able to predict, involuntarily earning for himself the reputation of a magician.

The table in Fig. 9 shows a layout of the two stories, as they were originally designed.

Going into a further level of detail, each act is divided into scenes, and each scene is divided into frames. At this level too the film is a "crab". Every sequence of frames was studied so that on the "opposite side" of the film there is a corresponding sequence in which the frames are similar, but mirrored and in the opposite order. Fig. 10 shows an example of a storyboard of two corresponding scenes.

These parallelisms are most visible at the centre of the film when the two characters (played by the same actor) meet in a dialogue that is outside of time and space.

| Euclid | Einstein |
|---|---|
| **First Part** | |
| 1 – Eclipse, Euclid is recognised as a seer, magician | 2 – Meeting of the Akademia Olympia, wife Mileva listens attentively. Bach plays on the violin, they cite Grossman and read Riemann, all wind up drunk |
| 3 – Lessons of Euclid on the beach where he presents his Geometry | 4 – Einstein goes to Besso's house from the patent office, strolling down the Kramgasse |
| 6 – Insistent questions by Apollonius about the parallel postulate and the existence of the gods, Euclid goes away confused | 5 – Einstein has a vision while improvising on the piano. He sees the whole world around him constituted of energy |
| **Second Part** | |
| 8 – Euclid returns to school, watches a propitiatory rite, apologises for having hurt someone's feelings, says that he needs a pause for reflection, and closes his academy the (Sunset). | 7 – Einstein tells his wife of his vision, who demands a mathematical exertion, otherwise neither she nor anyone else can understand it. |
| 10 – Apollonius, with his teacher's "method of proof", discovers that the sum of the angles of a triangle is two right angles, and even though the result is officially written on papyrus, he doesn't claim his own discovery, but instead writes Euclid's name. | 9 – Albert explains his vision to his friends philosophically. The group is given the name Akademia Olympia. His wife, rather than participating, gets up with an air of having had enough and abandons the group. |
| 11 – Archimedes goes to visit Euclid, tells him that his teachings have been very successful, and that even Apollonius had arrived at a new proof. "If you'll come back, we'll greet you with open arms". Euclid closes himself in the house and, after five weeks of trials, he is able to prove that the two statements about straight lines and triangles are equivalent. | 12 – Albert understands that his vision has to have a mathematical formulation. The first arrow of the dialogue. Einstein and Euclid meet outside of space-time. |

(the arrow flies past – the two characters talk to each other)

**Fig. 9.** The general scheme of the film *Assioma 5 – Oppositi e paralleli*

| Euclid | Einstein |
| --- | --- |
| 13 – Dialogue between Euclid and Einstein. An arrow after the conversation. Euclid comprehends that triangles are the same even when drawn on a curved surfaces, and the entire theory is founded on axioms. | 14 – Michele Besso goes to visit Albert. Einstein closes himself in his house for five weeks and falls asleep on the final formulation of the theory of relativity $E = mc^2$. Mileva reads the formula and is moved. |
| 16 – Euclid returns to the academy and sees Apollonius, who calls his academy "Olympia" because he believes in the gods. | 15 – Mileva is proud, gives one of her inventions the name of "Einstein". "Because we are one-stone". |
| 18 – Apollonius asks Euclid to stay with them. Euclid dictates the fifth axiom, without a proof: "Think of it as a constructive failure". | 17 – The whole Academy celebrates their graduation on the mountains, but it is sad because they will soon have to disband (dawn). |
| Third Part (the arrow reaches its target, Einstein's pendulum) | |
| 20 – Euclide has a vision. He sees the whole history of geometry that will unfold through the centuries. | 19 – Einstein dreams of Euclid and the arrow. He awakes, takes Riemann's book, and reads the fifth axiom. Albert mails his article to various astronomers. One reads its and, laughing, crumples it into a ball. A second reads it and, shaking his head, throws it away. |
| 21 – Euclid tells Archimedes, the rational person in the group, of his vision. Archimedes understands but sees that certain things cannot be part of his books. | 22 – Albert teaches his theory in the classroom with only three people in attendance: Besso, Chavan, Schenk. Colleagues at the patent office (1908). |
| 23 – Triumph at the Academy, all present celebrate the draft of thirteen papyri. A symposium where there is music, discussion, and drinking. Euclid is elected a kind of symposiarch. The scene concludes with Euclid spreading wine in honour of Dionysius. | 24 – A third astronomer receives the article, nods in approval, and does the calculations in time for the Eclipse. Recognition of the validity of Einstein's theories (1919). |

**Fig. 9.** (continued)

237

**Fig. 10.** Extract from a storyboard

## Akademia Olympia, an Open and Unanimous System

*Akademia Olympia* is the name that was given (recalling Albert Einstein and his friends) to the group of people who collaborated on the making of this film. We think that it might be interesting to explain the system by which our group was able to create something absolutely in common.

As opposed to a social organization, in which every decision is adopted with the law of the majority, our preferred criterion was that each decision be made on the basis of unanimous of consensus. Practically, this meant that every proposal had to be examined and approved with the consensus of all those present, on the condition that everyone was ready to justify and share his or her own suggestions and objections.

As a result, the general consensus was achieved by degrees, with the constant succession of suggestions and counter suggestions leading up to something that, in the end, everyone was able to agree on.

The *Akademia Olympia* was always a meeting place that was open to everyone: whoever found out about the project could decide to participate, could collaborate on a equal footing with everyone else, and had the same power to make decision. It is obvious that the two components, unanimous consensus and an open system, increased the capacity for self-criticism, allowed the arguments discussed to verified and clarified more efficiently, and furnished an indispensable fund of knowledge on the part of experts and technicians in various areas.

## Bibliography

[1]  Douglas R. Hofstadter (1979) *Gödel Escher Bach: The Eternal Golden Braid*, Pearson Education, London.
[2]  M.C. Escher (1992) *Grafici e Disegni*, Evergreen, Germany.
[3]  Dennis Overbye (2000) *Einstein in Love*, Bloomsbury, London.

# Mathematics and Wine

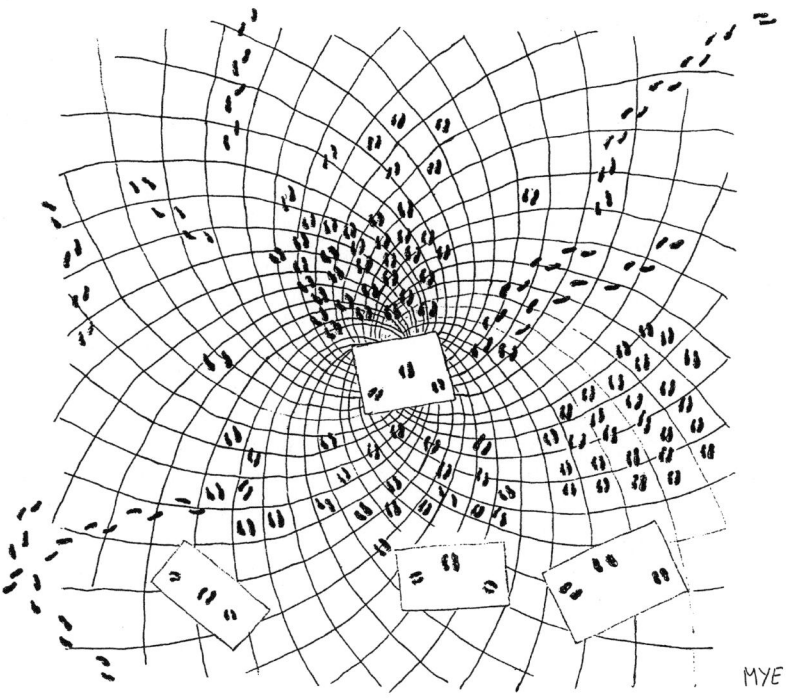

# Mathematics and Wine*

Antonio Terni

I have to confess that one of my dreams has come true: to be a speaker at a mathematics conference! For me, mathematics is simply a hobby: I don't understand much, but the little that I do understand fascinates me, and sometimes I am unexpectedly inspired in my work, which is cultivating vineyards, transforming grapes into wine, bottling it, and selling it around the world. My wine estate is in the area of the Conero Riviera, just south of Ancona, and looks out on same the Adriatic Sea which, here in Venice, mingles with the land in this extraordinary lagoon.

Wine producing is a genuinely fascinating profession, because it means turning nature's fruit into a product that carries throughout the world the memory of what the French call *terroir*: the interaction of soil, microclimate, grapevines, and the traditions of a special land.

Each *terroir* has a characteristic grapevine: Burgundy its Pinot Noir; Chianti its Sangiovese; Langhe its Nebbiolo, and so on. These grapevines, each in its own area, after decades and sometimes centuries of cultivation, are better able than any others, to exalt the pedoclimactic characteristics of that territory. The grapevine that is typical of the area of the Conero is the Montepulciano, and it is precisely with Montepulciano that the Rosso Conero is produced, which is the wine that best represents our company.

Sometimes, however, we get the urge to try to produce wines from other types of grapevines, to see what will happen, as well as to escape the monotony of years and years of Montepulciano. It was because of this that about ten years ago, since I had to replant a few hectares of Montepulciano, I decided to plant a few rows of Merlot and Syrah as well, two grapevines of French origin that easily adapt to other climates. In 1997 we harvested the first few hundredweights of Merlot and Syrah, and the problem immediately arose of what to do with them: on the one hand, we couldn't use them for the Rosso Conero because by law only a maximum of 15% of Sangiovese can be added to Montepulciano, and on the other hand, there were too few grapes to make individual wines of them. Together with Attilio Pagli, our

* Translated by Kim Williams

oenologist, we decided to produce a so-called blend, that is, to mix all of the Merlot and Syrah with equal quantities of Montepulciano, thus obtaining a totally new wine.

Each time that a wine producing company makes a new wine it begins with a drama: finding the right name. The name and the label make the first impression from the bottle of wine to the consumer, and we all know how important first impressions that we get of a person or a product are. Most of the time we try to find names that have some connection with our region, using words to evoke bouquets and tastes that are typical of this land, but we don't always manage to do it. Made-up names are almost always already registered by someone else in the world, and the names that are not already registered are dangerously similar to other registered names, running the risk of having someone contest the name when the wine is already on the market. Some producers, among them this present author, turns to their hobbies or extra-curricular activities, or to their own "visions". One of my friends, a great fan of Duke Ellington, called his wine "For Duke"; I myself, in honour of Bob Dylan, one of my favourite musicians, called one of my wines "Visions of J.", inspired by one of his most beautiful songs. Another producer gave his wine the name of "Where the dreams have no end...". So you can see that it is perfectly acceptable to use one's imagination, even in the most reckless ways.

And it is here that wine and mathematics met. One of the books that made the biggest impressions on me was *Chaos* by James Gleick. This is a book of decidedly popular science, which introduces the novice reader to complexity and to the unpredictability of a universe described by non-linear equations such that a miniscule difference in initial conditions can lead, over shorter or longer periods of time, to enormous differences in the later behaviour of a given phenomenon. The typical example is, in meteorology, the so-called butterfly effect, according to which the simple fluttering of a butterfly in Peking can cause, months later, a cyclone in the Caribbean.

The Mandelbrot set is the most fascinating of the graphic descriptions of chaos: the simplest iteration in the plane of complex numbers leads, depending on the choice of the initial numbers, to convergence or divergence, and the confines between points that lead to divergence and those leading to convergence is a fractal of infinite complexity. If we then attribute to the points in which the iteration diverges different colours in function of the speed of the divergence, we obtain figures that no artist would ever be able to even imagine. There exist many software programs that allow us to explore the Mandelbrot set, but the one that I like the best was developed by David E. Joyce of the Department of Mathematics and Computer Science of Clark University in Worchester, Massachusetts, which in long ago 1994 permitted entrance into the set by varying the various parameters. And further, with unusual generosity, the site developer allowed anyone who wanted to download the images generated and use them for whatever purpose he or she wanted.

When I connected to the Internet for the first time, in 1995 (I feel like I'm talking about prehistory), I had just read Gleick's book and naturally the word "Mandelbrot" was one of the first I looked up, bringing me almost immediately to the site that I just mentioned. As soon as I understood how the program worked, I began to explore enthusiastically the Mandelbrot set, making fascinating images, even if the slowness of transmission at that time meant that I had to wait for a half an hour at a time for

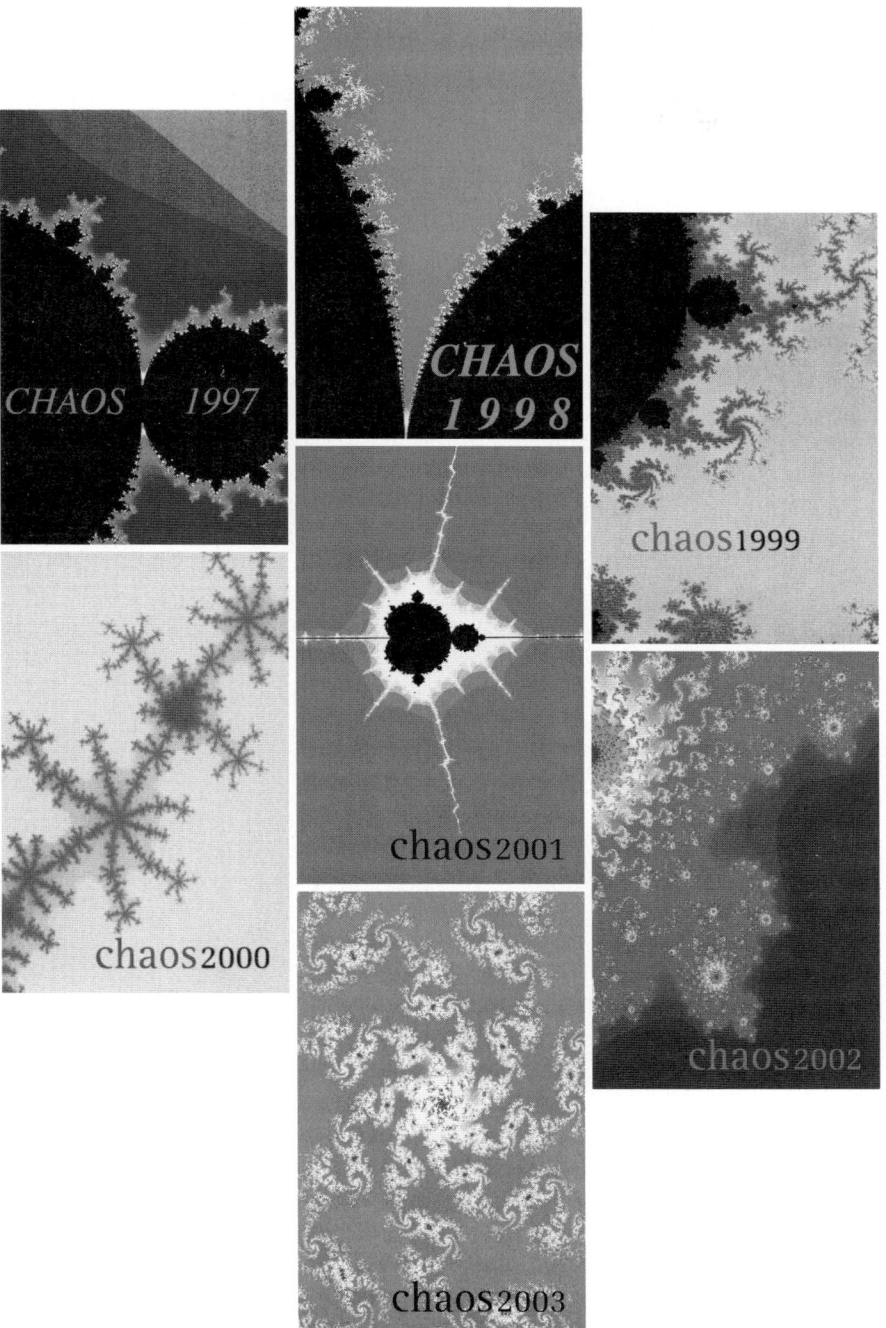

**Fig. 1.** (see the section in colour)

243

the result of my explorations. I showed some of the images to my oenologist, who, even though he is outstanding in his profession, harboured some of the terror that is typical in the face of mathematics and so refused to listen to my technical explanations. But he told me that those images, whose origins he refused to understand, would have been perfect for labels for wine. Obviously I remembered this when, two or three years later, I was searching for the name for my new wine and so it was natural for me not only to use the images from the Mandelbrot set, but to call the wine *Chaos*.

However, I had to find some justification for this name, if for no other reason than to be able to answer the inevitable questions from friends or clients. And so I recalled another thing that my oenologist told me, that is, how in a normal red wine there have been identified to date more than 2000 components. Some of these, such as water and alcohol, are present in large quantities, while others are present in infinitesimal quantities. But it is precisely these components (all kinds of acids, glycerine, polyphenols, antocyans, and so on) that give character to the wine, making a Lambrusco different from a Chianti and a Barolo different from another Barolo. And, given that the quality of the wine is apparently linked to the presence of these substances and their mutual interaction, it seemed to me that the theory of Chaos could be invoked to explain how the sensations that the wine produces in the person who tastes it cannot be in any way controllable in a precise and measurable relationship of cause and effect.

A summary of all this is found on the back of the label, which says:

*Chaos theory explains why certain patterns cannot be fully explained. And wine – this wine, any wine – cannot be explained by the countless interactions between its compounds. All the better.*

# Homage to Alfred Döblin
# and Vincent Doeblin

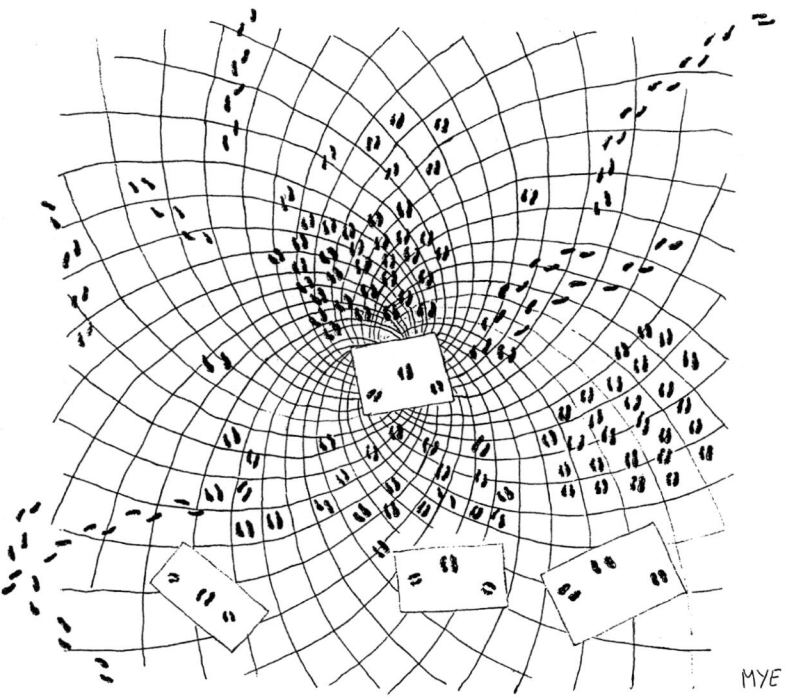

# Doeblin and Kolmogorov:
# The Mathematisation of Probability in the Thirties*

Carlo Boldrighini

My talk refers to the chapter by Marc Petit in this book, which tells the extraordinary story of Wolfgang Doeblin, son of the famous German writer Alfred Döblin[1], and his *pli cacheté*, the sealed envelope deposited by the young Doeblin with the *Académie des Sciences* in 1940, shortly before his tragic death. The biographies of the Döblins, father and son, have been narrated skillfully and attentively by Marc Petit in *L'équation de Kolmogoroff* [1]. The title of the note deposited by Wolfgang with the *Académie des Sciences* is precisely this: *L'équation de Kolmogoroff*.

My goal here is to illustrate Wolfgang Doeblin's contribution to probability theory, and in particular the relation between Doeblin and the Russian mathematician Andrei Nikolaevich Kolmogorov. It is Kolmogorov whose name appears in the note deposited with the *Académie des Sciences* and who can be considered the founder of modern probability.

Opened a few years ago, the contents of the sealed envelope revealed that Doeblin had been the first to obtain an understanding of some important results in the modern theory of stochastic processes. Doeblin's contribution to probability theory far exceeds "Doeblin's Theorem" on Markov chains, well known to any reader of probability textbooks, and has something prodigious to it, considering the brevity and the dramatic circumstances of his life.

Where possible, I will try to limit the technical aspects to make this text comprehensible to non-experts also. However, I cannot avoid introducing mathematical formulæ, but these should not provide difficulties to readers with some background in probability theory.

I will start with a short historical framing of probability, and then dwell upon the profound transformations mostly due to the work of Kolmogorov in the thirties - the years of Doeblin. Probability, in fact, "became" mathematics in those years, among

---

* *Translated by Sarah Wolf*
[1] *Once the Döblin family had obtained the French naturalisation in 1936, the young mathematician Wolfgang Döblin changed his name to Vincent Doblin. However, he continued to sign his work with his "true" name, using the Alsatian orthography Doeblin.*

vigorous objections coming not only from the so called *pure* mathematicians, but also from specialists of the sector. The discussion, in which Bruno de Finetti was one of the major protagonists, still cannot be considered completely settled and deserves a brief digression.

I will then outline Wolfgang Doeblin's scientific biography focusing on some salient points in the work of this unfortunate young man who, in just a few years developed from a student of Kolmogorov's work to one of his greatest emulators in the élite of experts in the sector.

## A Short History of Probability From the Weak Law of Large Numbers to the Strong Law of Large Numbers

Probability emerged in connection to gambling and its beginnings are connected to the names of Pacioli (1445–1514), Tartaglia (1499–1557), Pascal (1623–1662) and Fermat (1601–1665). This first period will not be dealt with here; rather, the problem of mathematisation shall be treated by considering just one example of central importance: the well known *law of large numbers*.

Let us begin with Jakob Bernoulli (1654–1705), who is the first to prove the law of large numbers in *Pars IV* of his *Ars Conjectandi* [2] released posthumously in 1713, which links probability and frequency in a precise manner. Bernoulli considers $n$ independent trials, each of which has two possible results, conventionally denoted by 1 ("success"), obtained with probability $p \in (0, 1)$, and 0 ("failure"), obtained with probability $q = 1 - p$. These models, referred to as "Bernoulli schemes" today, were and still are of fundamental importance for the development of ideas in probability theory and statistics. The most common example is given by the repeated throwing of a fair coin, the two results ("heads" and "tails") having the same probability $p = q = 1/2$.

Bernoulli proved that the number of sucesses in $n$ trials, $S_n = \sum_{k=1}^{n} \xi_k$, where $\xi_k \in \{0, 1\}$ are the results of the trials, is distributed according to the "binomial distribution": that is, it is a "random variable" which takes each value $k = 0, 1, \ldots, n$ with probability

$$P(S_n = k) = \binom{n}{k} p^k q^{n-k}, \quad \text{where} \quad \binom{n}{k} = \frac{n!}{k!(n-k)!} \, .$$

The name of this probability law is derived from the binomial coefficient $\binom{n}{k}$. The mean value (denoted by $M(.)$) of the random variable $S_n$ is easily seen to be $M(S_n) = np$. Bernoulli's law of large numbers now states that

$$\text{for all } \varepsilon \in (0, 1) \text{ we have} \quad \lim_{n \to \infty} P\left(\left\{\left|\frac{S_n}{n} - p\right| \leq \varepsilon\right\}\right) = 1 \,, \quad (1)$$

where $S_n/n$ is the "empirical frequency of success". The proof can nowadays be done in a single line, using "Chebyshev's inequality" which is true for any random variable $\xi$: $P(\{|\xi| \geq a\}) \leq M(\xi^2)/a^2$. Let $\xi = S_n - np$, then $M(\xi^2) = Var(S_n)$ is the variance of $S_n$,$^2$ and since $Var(S_n) = npq$, equ. (1) is obtained by letting $a = \varepsilon n$.

Today, this result is referred to as the "weak law of large numbers". It states that if the number of trials is very large, the empirical frequency is "almost certainly" close to the "a priori" probability $p$.

However, it does not say anything about the behaviour of the ratio $S_n/n$ in a specific case when the number of trials, $n$, grows. That is, it does not give information on the single "trajectories" of frequencies. We thus cannot even be sure that in every concrete case $S_n/n$ actually converges to the probability $p$ for $n \to \infty$.

For this last assertion, which today is called the *strong law of large numbers*, one must make use of a probability on the space of infinite sequences (infinite Bernoulli schemes). This becomes available only at the beginning of the twentieth century, when the modern theory of measure and integration appears, essentially by work of E. Borel (1871–1956) and his student H. Lebesgue (1875–1941).

The first result on the strong law of large numbers, owing to Borel, is in fact expressed not in probabilistic form, but as a property of the real numbers. Let us consider the real numbers in the interval $[0, 1]$, written to the base 2 instead of to the base 10. Hence, for each $x \in [0, 1]$ we will write $0.x_1 x_2 x_3 \ldots$, where the $x_k$ can take the values 0 or 1.

If we now let $S_n(x) = \sum_{k=1}^{n} x_k$, the quantity $S_n(x)/n$ is the frequency of the cifer 1 under the first $n$ cifers of the binary representation (i.e. to the base 2) of the number $x$.

Borel showed that this ratio tends to $1/2$ for all points in the interval, except for a set of points that has a total length of zero. In more precise terms, the following theorem holds.

**Theorem 1 (Borel, 1909).** *Excepting a set of points $x$ of Lebesgue measure zero,*

$$\lim_{n \to \infty} \frac{S_n(x)}{n} = \frac{1}{2}. \tag{2}$$

The probabilistic interpretation of this result is immediate. In fact, it is easy to see that the position $x_j = 0$, or $x_j = 1$, identifies $2^{j-1}$ intervals with a summed total length of $1/2$ for any $j$. Interpreting the length as probability and the $x_j$ as random variables, one can also verify the independence of the results for different values of $j$. Thus, an infinite sequence of throws of a fair coin is represented: we have constructed the probability of the infinite Bernoulli scheme with $p = 0.5$, and surprisingly, it coincides with the Lebesgue measure on the interval $[0, 1]$.

The result (2) is called the "strong law of large numbers".[3]

---

[2] *The variance of a random variable $\xi$ is defined as the average squared distance from the mean value: $Var(\xi) = M((\xi - M(\xi))^2)$.*

[3] *In the language of measure theory, the weak and strong laws of large numbers correspond to two different ways of interpreting the convergenge $S_n/n \to p$: it is interpreted in the sense of "convergence in measure" or "convergence in probability" for the weak law of large numbers and in the more restrictive sense of "almost sure convergence" (that is, except for a set of points with measure zero) for the strong law of large numbers.*

In the same year, 1909, Borel makes a first contribution to the general problem of the mathematisation of probability: in his article *Sur les probabilités dénombrables et leurs applications arithmétiques* [3], he indicates in fact the possibility to construct a mathematical theory of probability in terms of measure theory, a programme which will then be carried out by Kolmogorov.

Also for the strong law of large numbers of general character in probabilistic form we have to wait for Kolmogorov. His classical result from 1930 is the following.

**Theorem 2 (*Kolmogorov, 1930*).** *Let* $\xi_1, \xi_2, \ldots$ *be independent random variables with finite variance and let* $S_n = \sum_{j=1}^{n} \xi_j$. *Then, if the series* $\sum_{k=1}^{\infty} \frac{Var(\xi_k)}{k}$ *converges,*

$$\lim_{n \to \infty} \frac{S_n - M(S_n)}{n} = 0 \qquad almost\ everywhere. \qquad (3)$$

## Probabilistic Models at the Beginning of the Twentieth Century: Markov Chains, Random Walks, Brownian Motion, Markov Processes

Apart from the results on the relations between Lebesgue's measure theory and probability, of which Borel's theorem that we have just seen is the most evident example, there was a strong impulse for the mathematisation of probability in the sense of the theory of (parabolic) partial differential equations. It came from the study of models involving no longer sequences of independent random variables, but random variables with a simple form of dependence: Markov chains with a finite number of states, introduced by the Russian mathematician Markov in 1907.

Another very important example is the "random walk" on a lattice, for example on the lattice of integers, which is again a Markov chain, but now with an infinite state space (in fact, the number of reachable points in the lattice is infinite). This last model has a continuous version, Brownian motion, which had and still has a role of primary importance in probability, physics and other sciences.

Here, we will give a short description of these models, which are actually quite natural.

### Markov Chains (Discrete Time)

Markov chains constitute the simplest scheme of dependent random variables. A homogeneous Markov chain, with state space given by a finite or countable set of elements $E = \{e_1, e_2, \ldots\}$, is a sequence of random variables that take a value in $E$: $X(t) \in E$ for $t = 0, 1, 2, \ldots$ (where $t$ denotes "time").

The dependence is restricted by the "Markov property": for each choice of $s_1 < s_2 < \ldots < s_m < t$ it states that for the conditional probabilities

$$P(X(t) = e_k | X(s_1) = e_1, X(s_2) = e_2, \ldots, X(s_m) = e_j) = P(X(t) = e_k | X(s_m) = e_j). \qquad (4)$$

This means that once the preceding state $X(s_m) = e_j$ is known, one loses the dependence on those $X_s$ with $s < s_m$ ("short memory"). Therefore, it is sufficient to know the conditional probabilities for one step $p_{jk} = P(X(t+1) = e_k | X(t) = e_j)$, which, if the chain is "homogeneous" (in time), do not depend on $t$.

If the state space is finite, say with $n$ elements, $E = \{e_1, e_2, \ldots, e_n\}$, it suffices to determine an $n \times n$ matrix, called the "stochastic transition matrix":

$$P = \begin{pmatrix} p_{11} & \cdots & p_{1n} \\ \vdots & \cdots & \vdots \\ p_{n1} & \cdots & p_{nn} \end{pmatrix} \tag{5}$$

Since $p_{ij}$ is the probability of transiting from state $i$ to state $j$ and probabilities always sum up to 1, the sums over rows are $\sum_{j=1}^{n} p_{ij} = 1$.

The conditional probabilities for $r$ steps are given by the powers of $P$ (products of rows and columns), $P^r$, and have elements:

$$p_{ij}(r) := (P^r)_{ij} = \sum_{k=1}^{n} p_{ik}(r-1) \cdot p_{kj} = P(X(t+r) = e_j | X(t) = e_i) . \tag{6}$$

The Markov chain is defined by $P$ and an initial probability law $\mu^0 = (\mu_1^0, \mu_2^0, \ldots, \mu_n^0)$, where $\mu_j^0 = P(X(0) = e_j)$: the probabilities $P(X(t) = e_j) = \mu_j^{(t)}$ are given by the matrix product $\mu^{(0)} P^t$ (where the measures $\mu^{(t)}$, $t = 0, 1, \ldots$ are understood as row vectors, or matrices with one row and $n$ columns). Therefore,

$$\mu_j^{(t)} = (\mu^{(0)} P^t)_j = \sum_{i=1}^{n} \mu_i^{(0)} p_{ij}(t) .$$

## Random Walk

Already before Markov, however, random walks, which are a simple model of erratic movement, scattering or "diffusion", were studied for various practical applications. Random walks are one of the fundamental models in probability theory and have numerous applications. In mathematical physics they are the basic model for the phenomenon of diffusion, be it diffusion of particles, diffusion of heat or of something else.

In the simplest setting, that is, for a one-dimensional walk with jumps of unit length to the left or right, the model is defined by a Markov chain with the state space consisting of the integers $Z$. The transition probabilities for each $x \in Z$ are given by

$$P(X(t+1) = x + u | X(t) = x) = \frac{1}{2} \quad \text{for} \quad u \in \{+1, -1\} , \tag{7}$$

and are zero if $u \neq \pm 1$. The motion can be described as follows:
one throws a coin; if the result is heads one makes a step to the right, if it is tails, to the left. The state space is infinite and there is no invariant equilibrium measure. If

one starts from the origin, that is $X(0) = 0$, then $X(t)$ is the sum of $t$ independent variables: *jumps* or increments, which we will denote by $u_r$, $r = 1, 2, \ldots$. The $u_r$ can take the values $+1$ or $-1$, and have mean value zero, $M(u_r) = 0$, and variance one, $Var(u_r) = 1$. For the position of the random walk $X(t)$ at time $t$ we thus have

$$X(t) = \sum_{r=1}^{t} u_r \ .$$

One of the fundamental theorems of probability applies to this quantity; in this case it can be formulated as follows:

**Theorem 3 (Central Limit Theorem).** *For $t \to \infty$ and for any interval $(a, b)$,*

$$P\left( \frac{X(t)}{\sqrt{t}} \in (a, b) \right) \to \frac{1}{\sqrt{2\pi}} \int_a^b e^{-(x^2/2)} dx \ . \tag{8}$$

Note the normalisation factor $\sqrt{t}$, which shows that, as a consequence of the Central Limit Theorem, the square of the displacement $X^2(t)$ is of the order $t$.

It is worth observing that this property of the relation between time and displacement was pointed out, perhaps for the first time, by Bachélier [4] in 1900 in an article on the variations of prices at the French stock market (which might suggest that stock price variations are results of independent random contributions).

## Brownian Motion

However, it was particularly mathematical physics that developed the model of the random walk, especially in its continuous version. This takes the name "Brownian motion" from a phenomenon that is known since the seventeenth century and is easily observable under the microscope: the strange "dance" of semimacroscopic particles in suspension in a fluid (like grains of dust that one can see moving in the air in the sun). The phenomenon was described scientifically at the beginning of the nineteenth century by the botanist Brown, hence the name.

The basic idea behind the mathematical view on Brownian motion can be explained with a few words. In mathematical physics the so-called "macroscopic description" formalises particulate media as a continuum. This is in fact a scaling limit: the natural units for measuring space, time, mass, etc. within the macroscopic description are much larger than those of the corresponding microscopic variables (e.g. the average distance and time of interaction between molecules or the mass of a molecule); the limit for which one obtains a continuous description is reached by letting the ratio of the macro- and microscopic units of the various variables approach infinity. Changes of scale are linked to each other in a way that depends on the phenomenon under consideration.

If we want to determine the scaling limit for the random walk that we have seen above, we have to take into account that the square of the distance which has been covered is of the order of the time, that is, $X^2(t) \approx t$, and thus, if $M$ is the spatial scale, the temporal scale must be $M^2$ (this change of scale is called "diffusive"). The mathematical definition of Brownian motion is given as a limit of the random walk, precisely:

if $t$ denotes time ($t$ is now a real number, no longer an integer), then the standard Brownian motion at time $t$ is defined as the limit

$$b(t) = \lim_{M \to \infty} \frac{X([M^2 t])}{M} ,$$

where $[.]$ denotes the integer part of a real number: $[x] = \max\{n \in Z : n \le x\}$. Due to the Central Limit Theorem, $b(t)$ is a Gaussian random variable with mean zero and variance $t$. Its "probability density function" at a point $x$ on the real line, that is, intuitively, the probability that $b(t)$ assumes values in a little interval with center $x$, divided by the length of the interval, is

$$p(t, x) = \frac{e^{-(x^2/2t)}}{\sqrt{2\pi t}}$$

and this function satisfies, as is easily verified, the partial differential equation

$$\frac{\partial}{\partial t} p(t, x) = \frac{1}{2} \frac{\partial^2}{\partial x^2} p(t, x) \tag{9}$$

known as "heat equation".

The physico-mathematical theory of Brownian motion came into being in the years 1905–1906 most of all by work of Einstein [5] and Smoluchowski [6]. These works contributed to the mathematisation of probability in linking Brownian motion to partial differential equations such as the heat equation. They also played a crucial role in establishing the molecular theory of matter and led to the affirmation of the basic role of statistical mechanics (and thus of probabilistic concepts) with respect to thermodynamics. In particular, the statistical nature of the second law of thermodynamics was clarified.[4]

## Markov Processes. The Chapman-Kolmogorov Equation

One speaks of a Markov "process" instead of a Markov chain, if time, i.e. $t$, is continuous. Assuming, as for Markov chains, that the state space is finite (or countable), the transition probabilities from a state $e_i$ at time $s$ to a state $e_j$ at time $t > s$ will, if the process is homogeneous in time, be functions of the difference $t - s$:

$$p_{ij}(t - s) = P(X(t) = e_j | X(s) = e_i) . \tag{10a}$$

If we have a continuous state space, as in the case of Brownian motion seen above, for which the state space consists in the whole real line $R$, then the probability of

---

[4] *The experimental confirmation of the theory of Brownian motion yielded the french physicist J. Perrin a Nobel Prize in 1926.*

transiting to any given single point is zero and we have to specify in general the transition probability from a state $x$ at time $s$ to a set of states $A$ at time $t > s$:

$$P_{t-s}(x, A) = P(X(t) \in A | X(s) = x) \, . \tag{10b}$$

The so-called "Chapman-Kolmogorov" equation for the process, an equation for the transition probabilities, follows from the Markov property and the partition equation: for each intermediate time step $u$, and for $s < u < t$,

$$p_{ij}(t - s) = \sum_{k=1}^{n} p_{ik}(u - s) p_{kj}(t - u) \tag{11a}$$

in the discrete case, while in the continuous case the equation is

$$P_{t-s}(x, A) = \int_E P_{u-s}(x, dy) P_{t-u}(y, A) \, . \tag{11b}$$

Differentiating the Chapman-Kolmogorov equation with respect to $s$ or to $t$, one obtains equations called the "first Kolmogorov equation" (or "backward equation") and the "second Kolmogorov equation" (or "forward equation"). For example, in the case of Brownian motion, one obtains the heat equation (9).

## Kolmogorov's Contribution: Axiomatisation, "Analytical Methods" and Markov Processes

Andrei N. Kolmogorov (1903–1987) had extensive interests and made fundamental contributions to various sectors of mathematics and also to physics. He was a student of one of the greatest Russian mathematicians, N.N. Lusin, known primarily for his results in the theory of functions of one real variable. Thanks to this school, Kolmogorov came to know the work of Borel and Lebesgue. His first works in probability go back to the middle of the 1920's and were written in collaboration with A. Khinchin, who also was a student of Lusin's. Khinchin had already obtained relevant results, in particular, he had proved the well-known "law of the iterated logarithm" (1924), which states the speed of convergence in Borel's strong law of large numbers more precisely.

Already in his first works Kolmogorov obtained important results, such as the convergence conditions for series of independent random variables. He began to concern himself with the foundations of probability at the end of the 1920's. In 1933, his most famous work appeared under the title *Grundbegriffe der Wahrscheinlichkeitsrechnung* (Foundations of the Theory of Probability [7]), published by Springer in Berlin. It contains the axiomatisation of probability theory, which hence became a branch of mathematics.

Here, Kolmogorov implemented the fundamental idea that mathematical probability should be based on general measure theory, anticipated not only by Borel, as we have seen, but also by other mathematicians, for example the Russian probabilist Bernstein. Two of the principal aspects in Kolmogorov's works of 1933 are particularly remarkable: the construction of probability distributions in infinite dimensional spaces in terms of finite dimensional distributions, which has allowed for the development of the theory of stochastic processes, and the general mathematical theory of conditional expectation.

Discussing the motivation of his work, Kolmogorov observes in the introduction to *Grundbegriffe*, referring directly to works on Brownian motion, that "these new problems arose of necessity, from some perfectly concrete physical problems".

Another very important contribution by Kolmogorov, titled "On analytical methods in probability theory", appeared in 1931 in *Mathematische Annalen* [8]. This work constitutes the foundations of the theory of Markov processes; the main object of study is the Chapman-Kolmogorov equation for transition probabilities of a Markov process, that we have seen above (equations (11a) and (11b)).

Kolmogorov did not study the realisations (the single trajectories) of the process $X(t)$ directly - Doeblin should be the first to do this in his *pli cacheté* - but derived from the Chapman-Kolmogorov equation those partial differential equations known today as the backward and the forward Kolmogorov equations, which I mentioned in the preceding section.

## A Fundamental Question: Is Probability Mathematics?

The mathematisation of probability in those years encountered strong philosophical and ideological difficulties. In fact, probability was generally not considered to be mathematics, and many people thought it was only partly possible to mathematise probability. The debate was naturally influenced by the fact that probability does not seem to be in line with the determinism which inspired the vision of mathematised science. This ideological problem was also faced by Kolmogorov in the Soviet Union. The positivist philosopher A. Comte was very critical of probability; however, also scientists who were well aware of the importance of probability, such as the renowned French mathematician Henry Poincaré and in part even Borel (who by the way can be considered the initiator of modern probability theory in France) did not think that probability was only mathematics. This opinion was based primarily on the fact that they considered some procedure of valuation and estimation an essential component of the discipline.

In the tradition of Pascal and Laplace, in fact, probability deals with uncertainty, with estimates connected to random events, and its laws constitute the "logic of uncertainty". There is no doubt that the conditions of uncertainty are a salient aspect of many practical applications, not only in gambling: it suffices to think of applications in insurance problems, which were already well developed at the end of the nineteenth century.

In reality, in many cases of practical interest, one makes estimates in conditions in which it is not at all clear whether or not "true probabilities" exist, of which these estimates are approximations. And in fact, the central philosophical problem, which has been debated for a long time, is precisely that of establishing what probabilities are and how to compute them, that is, how to extract numbers from conditions of

255

uncertainty. In some languages the name of the discipline, involving "probability" and "calculation"[5], still bears witness of the historic importance of the question.

The calculation seems well-defined only in the case of the "classical model", where the sample space $E$ has a finite number $|E| = n$ of points, all of which are equivalent, such that it is natural to assign probability $1/n$ to each point. The probability of an arbitrary event $A$, that is, an arbitrary subset of the sample space, $A \subseteq E$, is thus given by the classical formula

$$P(A) = \frac{\text{number of favourable cases}}{\text{number of possible cases}} = \frac{|A|}{n} \, .$$

There have been attempts to reduce probability theory to the classical formula, but, apart from the difficulty to establish when one can assume that the possible results are "equivalent", it is evident that in many cases this formula cannot be applied. If, for example, we want to find the probability that a marksman hits the target under certain circumstances, we cannot suppose that the shots are uniformly distributed in a certain area around the mark.

One solution, suggested by the law of large numbers and preferred by scientist of positivist attitude, is the frequentist approach which identifies the probability of an event with its frequency in a sequence of independent trials. However, this approach is not free of problems either. The main difficulty is that the law of large numbers provides a limit when the number of trials tends to infinity. Therefore it requires repeatable events. And even for those there is a problem: if one stops at a finite number of experiments, it is not entirely correct to say that the frequency one obtains is an approximation of the "true probability" in the same sense as in a measurement of a length with a certain degree of precision. In fact, as we have seen, the frequency is close to the probability only *with a certain probability* which again depends on the "true probability".

A radical solution to the problem of the nature of probability was put forth by one of the principal probabilists of the 20th century, the Italian Bruno de Finetti [9]: it is to consider "subjective probability". The idea consists in giving up the concept of "true probability" and in considering probability simply as a subjective estimate of uncertain cases, given according to certain rules. An estimate can be improved by corrections that are obtained on the base of repeated experiments which provide "a posteriori" probabilities. If the experiment can be iterated an infinite number of times, the frequentist approach can be justified.

When using subjective probability, it seems possible to assign probabilities also to events which are not repeatable, or repeatable only a few times, as for example in economic or political decisions etc.

De Finetti's ideas have influenced the development of statistics, however, as far as probability is concerned, for various reasons they present the disadvantage of re-

---

[5] *Translator's note: For example in Italian, "calcolo delle probabilità" literally means "calculation of probabilities". The same is true of the French "Calcul de Probabilités" and the German "Wahrscheinlichkeitsrechnung". However, similar expressions do not exists in English and Russian.*

quiring a mathematical approach which leads to considerable complications. Kolmogorov's axiomatisation on the other hand proceeds in a completely different manner: it eliminates the problem of calculating probabilities by considering these given right from the beginning. Thus, if the sample space $E$ is discrete, probabilities are assigned to the single points of $E$, while in the general case, for example if $E$ is continuous, a finite measure in the sense of Lebesgue's measure theory is assigned to a certain class of subsets of $E$ and normalised in such a way that the total mass is equal to 1. Probability theory can therefore make use of the powerful machinery of the Lebesgue measure.

The axiomatisation has met various objections, and not only for having eliminated the problem of calculating probabilities. One can actually think that axiomatised probability is a part of measure theory, that is, a part of mathematical analysis, and this is still the opinion of quite a few mathematicians, though mostly remote from the sector. For example, in the book of M. Kline [10], one of the most widely-circulated textbooks on the history of mathematics, there is no trace of probability in the volume dedicated to modern mathematics (from 1700 onwards). Kolmogorov answered to this type of objections that probability differs from measure theory in its intuitive-conceptual dimension, which is not irrelevant in that it determines how problems are posed, and which to a great extent derives from the old "logic of uncertainty". For example, the concept of "independence", fundamental in probability, does not have a particular sense in an analytical context.

What, however, is probability for Kolmogorov? The entry "Probability, *mathematical*" written by him for the Encyclopaedia of Mathematics [11] illustrates:

> A numerical characteristic expressing the degree to which some given event is likely to occur under certain given conditions which may recur an unlimited number of times.

The entry refers to both uncertainty (*likely to occur*) and the infinite sequences of the frequentist definition. This aspect is again underlined later on: "the concept of 'probability' describes a special type of connection between the phenomena, which are typical of mass processes". The question what probabilities are and how they are calculated is left open, or better, is placed inside something that is similar to a platonic archetype. Probabilities can sometimes be calculated on the base of the classical definition, or one can call for the statistical approach, and they can also be given a priori in the axiomatic approach. However, Kolmogorov says:

> Neither these axioms, nor the classical approach to probability nor the statistical approach fully explains the real meaning of the concept of 'probability'; they are merely approximations to a more and more complete description.

Subjective probability seems unacceptable to Kolmogorov. This opinion may however be partly induced by his position as a Soviet citizen. One cannot, he says, attribute a probability to all events. "The assumption that a definite probability [...] in fact exists for a given event under given conditions is a *hypothesis* which must be verified or justified in each individual case". What can be said about the probability of single events? For example, if one asks what the weather will be like in Rome on the 15th of August in 2010, one can confidently say that the weather will probably be nice. The reliability of the answer is in reality based on objective climatic regularities. This use of probability, according to Kolmogorov, does not justify the subjective approach by de Finetti, that is described in the Encyclopaedia entry in a rather exaggerated way:

> Accordingly, the calculation of mathematical probability in order to arrive at a degree of reliability of certain statements concerning individual events is no longer a mere expression of the subjective belief that the event will or will not take place. Such an idealistic, subjective understanding of the sense of mathematical probability is erroneous. If pursued to its logical conclusion, it would result in the absurd claim that valid conclusions about the world around us can be arrived at in complete ignorance, by merely analyzing subjective, more or less reliable opinions.

For a discussion of the foundations of probability see also the chapter by Fabio Spizzichino in this book.

## Doeblin, Markov Chains and the Kolmogorov Equation

Despite his short life, Wolfgang Doeblin (1915–1940) has played a rather important role in the development of the potential offered by Kolmogorov's new approach and in spreading the new ideas in the Western world. His fundamental contributions concern Markov chains and Markov processes.

Doeblin begins his activities on probability in Paris mostly under the supervision of Paul Lévy and Maurice Fréchet. His interest in Markov chains, to which his renown is mainly connected, owes to Fréchet, the only specialist on this topic in France at the time. Fréchet was in direct contact with Kolmogorov, who often mentioned that during his visit to France in 1930-1931 he spent entire days with Fréchet discussing precisely about Markov chains. At the request of Fréchet, Doeblin contributed also to the diffusion of Kolmogorov's new ideas by translating *Grundbegriffe* into French.

Studying Markov chains, Doeblin made great progress in a short time, arriving in 1936, at only 20 years of age, at the fundamental theorem of ergodicity which now carries his name. For the proof he created a very elegant method, nowadays called *coupling*, that was further developed in recent times in the context of the theory of Markov processes. Doeblin's work [12] was not published in a prestigious journal, but it appeared in the Review of Mathematics of the Interbalcanic Union, that had a very brief lifetime, cut short by the war.

I think it is worthwhile presenting a brief sketch of the proof of Doeblin's theorem, to give the possibility of understanding the basic idea of the *coupling* construction, which, as is frequently the case, is simple and brilliant at the same time. Let us therefore return to Markov chains with a finite number of states, given by a stochastic transition matrix as in (5) and an initial probability measure. The theorem requires only one simple definition, namely that of an ergodic chain.

**Definition 1.** *A Markov chain is ergodic if for some integer $r > 0$ the matrix $P^r$ has only positive matrix elements (as in (6)): $p_{ij}(r) > 0$.*

**Theorem 4 (Doeblin's Theorem).** *Given an ergodic chain, there is a unique probability measure $\pi = \{\pi_i : i = 1, \ldots, n\}$ such that for an arbitrary initial state $e_j$, for $t \to \infty$:*

$$p_{ji}(t) \to \pi_i .$$

Practically, whatever the initial situation may be, the theorem confirms the convergence of the chain to a unique final "equilibrium" (or "stationary") probability, the probability $\pi$.

In the proof, an essential step consists in seeing that, however two initial states $e_j, e_k$ and a final state are chosen,

$$\lim_{t \to \infty} (p_{ji}(t) - p_{ki}(t)) = 0 . \tag{12}$$

The statement (12) can easily be obtained via the *coupling* construction, which is an appropriate joint distribution of two copies of the same Markov chain, $X(t)$ and $Y(t)$, with initial states $e_j$ and $e_k$ respectively (assume initial probabilities which assign the starting points $e_j$ and $e_k$ with certainty, so that the probabilities at time $t$ are $\{p_{ji}(t) : i = 1, \ldots, n\}$ and $\{p_{ki}(t) : i = 1, \ldots, n\}$). Let $X(t)$ and $Y(t)$ proceed as two independent chains up to the point where they first meet, that is, up to the first time $t$ such that $X(t) = Y(t)$. Denoting this first meeting time, which is a random variable, by $T$, one then assumes for $t > T$ that the two chains proceed as one single chain: $X(t) = Y(t)$ for $t > T$. This defines the *coupling* of the two chains.

The single chains (more precisely, their marginal distributions) always behave like the original chain, therefore

$$p_{ji}(t) = P(X(t) = e_i | X(0) = e_j), \quad p_{ki} = P(Y(t) = e_i | Y(0) = e_k) .$$

By definition, the two chains can differ only as long as time $T$ has not yet been reached: $P(X(t) \neq Y(t)) = P(t < T)$. Thus, considering the conditional probabilities under the condition $t > T$, we have equality:

$$P(X(t) = e_i | X(0) = e_j, t \geq T) = P(Y(t) = e_i | Y(0) = e_k, t \geq T) .$$

If the probability $P(t \geq T)$ tends to 1 for $t \to \infty$, the condition disappears, the conditional probabilities tend to $p_{ji}(t)$, respectively $p_{ki}(t)$ and under the given assumption (12) is proved.

It is immediately seen that if $P^r$ has only positive elements, the same is true for all larger powers $P^s$ with $s > r$. Suppose for simplicity that $r = 1$, i.e., the matrix $P$ itself has only strictly positive elements, and let $a$ be their minimum. Then the probability that the two chains meet is at least $a$ at each step and hence $P(t \leq T) \leq (1 - a)^t$, which tends to zero for $t \to \infty$, and (12) is proved.

The proof of (12) for $r > 1$ requires small modifications. The proof of the theorem is completed by showing that the sequences $p_{ji}(t)$ are Cauchy sequences.

259

Doeblin's results are certainly not restricted to the famous theorem. He has produced important contributions to the theory of Markov chains with infinite state space, for which he discovered a property called "recurrence", which is fundamental for understanding their behaviour. He obtained remarkable results in the study of domains of attraction for sums of independent random variables (series schemes), one of the major research fields of P. Lévy, who was a great French probabilist and a teacher of Doeblin.

The scientific life of Wolfgang Doeblin does not even cease when, in November 1938, by his own choice he becomes a regular soldier in the French army, under the frenchified name of Vincent Doblin. He manages to work under difficult conditions, at night, in the little free time and during leaves. Many of his notes do not contain detailed proofs.

Towards the end of his life, at the beginning of the Second World War, he worked on Markov processes, studying in particular the Chapman-Kolmogorov equation. He was the first to tackle the study of the process "trajectory-wise". Perhaps feeling that he had discovered something fundamental and wanting to leave a trace of it, Wolfgang Doeblin managed to complete a manuscript and to deposit it with the *Académie des Sciences* shortly before his tragic end, as *pli cacheté* in the appropriate archive, addressed to posterity.

## Bibliography

[1]  M. Pétit. *L'Equation de Kolmogoroff.* Ramsay, 2003.

[2]  J. Bernoulli. *Ars Conjectandi.* Basel, 1713.

[3]  E. Borel. Sur les probabilités dénombrables et leurs applications arithmétiques. *Rendiconti del Circolo Matematico di Palermo,* 26:247–271, 1909.

[4]  L. Bachélier. Théorie de la Spéculation. *Annales Scientifiques de l' École Normale Supérieure,* 17:21–86, 1900.

[5]  A. Einstein. Über die von der molekularkinetischen Theorie der Wärme geforderte Bewegung von in ruhenden Flüssigkeiten suspendierten Teilchen. *Annalen der Physik,* 17:549–560, 1905.

[6]  M. Smoluchowski. Zur kinetischen Theorie der Brownschen Molekularbewegung und der Supensionen. *Annalen der Physik,* 21:756–780, 1905.

[7]  A.N. Kolmogorov. *Grundbegriffe der Wahrscheinlichkeitsrechnung.* Springer-Verlag, Berlin, 1933. English translation: *Foundations of the Theory of Probability.* Chelsea Publishing Company, 1956.

[8]  A.N. Kolmogorov. Über die analytischen Methoden in der Wahrscheinlichkeitsrechnung. *Mathematische Annalen,* 104:415–458, 1931.

[9]  B. de Finetti, editor. *La logica dell'incerto.* Il Saggiatore, 1989.

[10]  M. Kline. *Mathematical Thought From Ancient to Modern Times.* Oxford University Press, 1972.

[11]  A.N. Kolmogorov. Entry: Probability, *mathematical.* In *Encyclopedia of Mathematics.* Kluwer Academic Publishers, Dordrecht, 1991. English translation of: Matematicheskaya Enziklopedia, vol. 1, Sovietskaya Enziklopedia, Moscow, 1977.

[12]  W. Doeblin. Exposé de la théorie des chaînes simples constantes de Markoff à un nombre fini d'états. *Revue de Mathématique de l'Union Interbalkanique,* 2:77–105, 1938.

# Wolfgang Doeblin and the Kolmogoroff Equation*

Marc Petit

On 18 May 2000, in the Archives Room of the Académie des sciences, 23, quai de Conti, Paris, the members of the Sealed Documents Committee began opening envelope number 11,668 entitled *Kolmogoroff's Equation*[1]. Inside was a school textbook from the "Docks ardennais" in the series 'Villes et paysages de France' [Urban and Rural Landscapes of France] and on its mauve cover a picture of the "rock of Bonnevie" in the Auvergne. Its pages were covered in fine spindly handwriting in blue-black ink; some pages were completely crossed out, others had become detached but lacked page numbers, indicative of the haste in which they had been written. The author of this paper, Wolfgang Doeblin, died aged 25 on 21 June 1940, just one day before the armistice. A telephonist with the 291st infantry regiment, he had hidden in a barn in a little village in the Vosges, and put a gun to his head so as not to be taken prisoner by the German army. Five months earlier, in February 1940, he had sent the Académie the mysterious textbook containing the results of his latest research in the field of probability.

Wolfgang Doeblin was not unknown in the world of mathematics. But France, his adoptive country, was taking its time acknowledging him, in line with the general disaffection in which his discipline was held in the 1950s, when Bourbaki mathematics reigned supreme. Yet, in Soviet Russia, in Australia and especially in the United States, other researchers, led by Joseph L. Doob and Kai Lai Chung, a probabilist of Chinese origin, recognised the importance of the young mathematician's work, in particular in the general theory of Markov chains. These could be used to create a model for a chain of events where the future depended only on the present and not on the past. Yet no-one had any idea of the contents of the sealed document, discovered by science historian Bernard Bru while working on correspondence between Wolfgang and his teacher (and former thesis supervisor), Maurice Fréchet, a brilliant prescient to the modern theory of probability, anticipating in certain respects the approach of Kiyosi Itô, Japanese founder of stochastic calculus (1944), but also

---

* *Translated by Anne-Marie Kerr*

[1] 'Kolmogoroff' was the usual spelling at this time. Nowadays, it would normally be written with a 'v'.

the Dubins-Schwarz, Yamada and Yamada-Ogura theorems (1965, 1973 and 1981, respectively). As early as 1947, the young mathematician's other mentor and leader of the French school of probability, Paul Lévy, correctly placed Wolfgang Doeblin on a par with Évariste Galois and Niels Henryk Abel, legendary figures in their field and comparable to Arthur Rimbaud, for both their precocious genius and meteoric rise to fame.

We must not, however, be tempted to romanticise this figure of Wolfgang Doeblin: he was no firebrand, he spoke little and pursued his passion in silence; he was also a workaholic, intuitive but with a firm grasp on reality, he was after all a soldier. Thirteen papers, as many reports, a doctoral thesis, several unedited works, including the celebrated sealed document, all written in under five years between 1936 and 1940 – such prolificness defies the imagination and represents an almost inhuman rate of creativity, especially when we consider the dramatic circumstances in which the young Jewish German intellectual managed to complete his work, emigrating to Zurich first, then to Paris in 1933.

**Fig. 1.** Wolfgang Doeblin (1915–1940).
Courtesy Stephan Doblin

Wolfgang was born in Berlin on 17 March 1915, the second [of 4] son of the great writer Alfred Döblin – author of, among others, *Berlin Alexanderplatz*, an iconic work of the modern age. Wolfgang's relationship with his father was an ambiguous one: open hostility coupled with a secret affinity. He was educated at the Protestant college of Königstadt and sat his Baccalaureate exams in the spring of 1933 in Berlin which was already under Nazi rule. A confirmed Marxist, he first considered studying political economy, then changed to mathematics, via statistics.

Along with Göttingen and Moscow, Paris was one of the very early centres for new mathematics in the 1930s, notably in the field of probability. The shadow of Henri Poincaré fell heavily on the brand new buildings of the institute bearing his name and dedicated to training top quality researchers in mathematics and mathematical physics. At its head, Émile Borel who had left cutting edge research to become a politician, and working with him, Georges Darmois, Arnaud Denjoy and Maurice Fréchet, among others, who would figure to various degrees in Wolfgang Doeblin's life and studies. Doeblin was admitted to the *Société mathématique de France* (French Mathematical Society) in the autumn of 1935. Under the supervision of Maurice Fréchet, he proposed a subject for his thesis on Markov chains: *Asymptotic properties of movements of certain types of simple chains*. Six months later, in June 1936, he had already gathered most of his results. His thesis was published in two instalments in Bucharest in 1937–1938. A first paper *The non-continuous case of chain probability* had been published earlier in the annals of Masaryk University, Brno. The speed with which Wolfgang progressed from student to researcher and even "master" is breathtaking. In 1935, he was studying the writings of Kolmogoroff, published in German, and translated *das Ergodenprinzip* (the Ergodic principle) as *the Ergoden principle*, demonstrating that he knew nothing on the subject. Yet, within a year, Wolfgang had reconstructed in his head the chain theory in its entirety, and would soon be going beyond Kolmogoroff's results and applying it generally in his paper on the *General theory of simple Markoff chains* published in 1940 in the *Annales scientifiques de l'École normale supérieure.*

Meanwhile Wolfgang had met Paul Lévy, the great mathematician, and certainly, with Wolfgang, the most original inventor of the time in matters of probability. From the outset, Lévy considered Doeblin a colleague rather than a disciple. Wolfgang found Lévy's work on the detailed study of Brownian movement stimulating reading, along with the writings of Andrei Nikolaevitch Kolmogoroff, whose *Foundations of the theory of probability* (1933) finally provided an axiomatic approach to the subject, and therefore the proper mathematical status which it seemed to have lacked. Moreover, at the Henri-Poincaré Institute and Jacques Hadamard's seminars at the Collège de France, Wolfgang Doeblin worked closely with some of the best French or French-speaking probabilists of his generation, though this closeness did not extend to his social life: Robert Fortet, Michel Loève and Jean Ville, who in his thesis of 1939 introduced the idea of the martingale for which Wolfgang, in his sealed document, found undreamt-of applications.

In October 1936 Wolfgang Doeblin became naturalised as a French citizen, together with his whole family – except his brother Peter who had emigrated to the United States – and changed his name to Vincent Doblin. Nevertheless he continued to autograph all his work in his "real" name, using the Alsacian spelling of Doeblin. He watched in anger and distress as the threat of Hitler's rule became more real with every passing month. With his Pomeranian Jewish origins, and as the son of an activist author whose name already figured on the Gestapo's hit list, Doeblin had remained privately loyal to his revolutionary beliefs. But now, according to Paul

Lévy, he broke his silence and spoke out against the all-pervading atmosphere of pacifism: "I have every right to voice my opinion" he exclaimed at the time of the Munich Agreement, "for I am one of those who know how to die for their beliefs." Refusing the privileges available to him, Wolfgang, a doctor of science, rejected the opportunity of attending Officers' Academy. He was drafted instead at Givet in the Ardennes as a simple soldier and was later moved to Sécheval when war was declared, then to Athienville, and finally to Oermingen on the frontline, not far from the Sarre border.

This is where, in his military quarters or telephonist's cabin, in the worst possible moral and material conditions, between November 1938 and May 1940, Wolfgang drafted the last of his mathematical output. At Givet, where he did his training, he combated depression by returning to research he had begun long before but never concluded, on the sum of independent variables. Entitled *The set of powers of one law of probability*, the article he sent in July 1939 to Steinhaus and Banach for publication in the Lvov review *Studia Mathematica* contains perhaps the most surprising result of the whole theory, known as 'universal laws': these are purely mathematical processes of selecting numbers at random, which, when repeated many times, lead to the creation not only of all limit laws, including the 'normal law' – the bell curve – but also of all 'infinitely divisible laws'. The remainder of the paper which relates to 'domains of partial attraction' is still undeveloped to this day. By his own admission, the author encrypted his drafts of the 'last theorem of Givet' which he sent to his brother Peter in Philadelphia. Perhaps one day someone will manage to decipher this mysterious code.

After some interruption to his work, Wolfgang returned to writing at Sécheval in the last few months of 1939, during the 'phoney war'. On night duty in his telephonist's cabin, he would draft, then re-write the material for the sealed document: *Research on the Chapman-Kolmogoroff equation*. In the summer of 1938 while walking on his own in the Alps, he had already envisaged the principal results of this work (see box). An unpublished notebook, found in a cupboard in the science faculty at Jussieu in December 2002, gives us an example of Doeblin's budding mathematical inventiveness. In Athienville, a tiny village in the Lorraine, Wolfgang completed his paper which he then sent to the Académie des sciences on 19 February 1940. The envelope was registered on 26 February. It lay completely forgotten until Bernard Bru rediscovered the trail of the missing manuscript while investigating the extraordinary career of Wolfgang Doeblin.

On 9 May, the German army advanced. Wolfgang knew he had nothing to lose: "I am a Jew", he had once said to the peasant who brought him a bowl of milk every night in Sécheval; "I always carry one shot with me, I won't let them take me." Over the following weeks, he fought heroically. On 19 May, he was mentioned in despatches at regiment level and awarded the *croix de guerre* (war cross). In 1945, he was mentioned a second time, posthumously, and awarded the *médaille militaire* (military medal) in 1948. Under repeated attack from the German army, Wolfgang's decimated battalion retreated ever further, from the Sarre border to the heart of the Vosges. The soldiers in his company were preparing to surrender. The armistice was only two days away, to be signed on 22 June.

**Fig. 2.** The textbook in which Wolfgang Doeblin hastily noted down the results of his latest research. Photo by Marc Petit. *Académie des sciences*, Paris

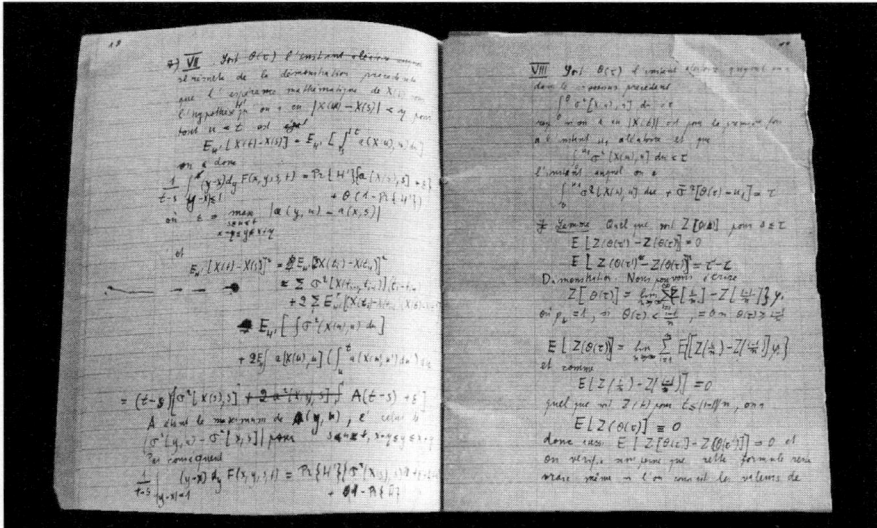

**Fig. 3.** Two pages of Wolfgang Doeblin's monography *Sur l'équation de Kolmogoroff*. Photo by Marc Petit, *Académie des sciences*, Paris

On the night of 20th to 21st, Wolfgang disappeared. Perhaps he had hoped to be able to continue hiding, or get behind enemy lines. He walked for a long time in the rain, came upon a small village, Housseras, on the edge of the forest, and sought shelter in a barn. In the morning, the Germans had surrounded the village. Wolfgang walked into the kitchen of the adjacent farmhouse, burned his identity papers on the fire, and, without a word, returned to his hiding place. Then a single shot was heard. No-one would know until April 1944 the identity of the soldier found dead in the hay. After searching in vain, Monette Tonnelat, a college friend and seemingly the woman he loved, was first to hear the news of his death. 'Vincent''s parents never fully recovered from the shock. When they died in 1957, they were buried in the same grave as their son in the little cemetery at Housseras. As for the sealed envelope, it would lie for almost another 50 years in one of the green boxes piled high in the Institute's archives, before a benevolent hand saved it from oblivion. Just like a message in a bottle washed ashore from a wreck, or a sunken treasure, returned to the world of the living.

**Fig. 4.** The archive room in the library of the *Académie des sciences* (Paris). Photo by Marc Petit

## Explanation – The Contents of the Envelope

It is often the case – and probably quite natural – that when some remarkable work has remained hidden and finally achieves some kind of recognition, it is also awarded virtues and attributes as exaggerated as its earlier lack of recognition was undeserved.

We have noticed a similar tendency with regard to Wolfgang Doeblin's sealed envelope number 11,668, so it would seem pertinent to indicate what the envelope does and does not contain.

Firstly, it does not contain the invention, nor even the draft construction of the concept of the stochastic integral, which has been attributed to K. Itô since 1944 (this integral is the cornerstone of all stochastic calculus). However, it does contain a trajectory analysis of Kolmogoroff's unidimensional equation linked to a diffusion coefficient and a derivation coefficient, where the fundamental martingales linked to this equation are highlighted. These martingales are also described as time-variant Brownian motion. This approach and these results were very new for the period: for example the notion of the martingale had only just been determined by Ville, a co-disciple of Wolfgang's at the Poincaré Institute. In this and other matters, Wolfgang Doeblin was approximately 20 years ahead of his time!

Finally, the envelope represents a hyphen or bridge between analytical research on Kolmogoroff's equation (pre-1940) and the establishment (post-1944) of diffusion processes and trajectory processes which would be largely advanced by Itô's research. It should also be noted that the new results attributable to Doeblin are not limited to those in the sealed document. He is also responsible for a detailed study of Markov chains and the invention of the notion of coupling between processes, which also figures partly in the document.

*Bernard Bru*, professor emeritus, René-Descartes University (Paris V)
*Marc Yor*, corresponding member of the Académie des sciences, professor at Pierre-et-Marie-Curie University (Paris VI)

## Further Information

Paul Lévy, *Revue d'histoire des sciences*, p 107, 1955

*Bulletin des sciences mathématiques*, 80, 60, 1956

*Matapli*, 68, 78, 2002

*Doeblin and Modern Probability*, Harry Cohn ed., American Mathematical Society, Providence (USA), 1993

W. Doeblin, *Comptes rendus de l'Académie des sciences* I, 331, 2000 (contains, other than the text of the sealed document, several related works by W Doeblin, an historical introduction with commentary by Bernard Bru and Marc Yor, and the complete bibliography of W Doeblin).

Bernard Bru and Marc Yor, *La Lettre de l'Académie des sciences*, 2, 16, 2001.

Marc Petit, *L'Equation de Kolmogoroff. Vie et mort de Wolfgang Doeblin, un génie dans la tourmente nazie*. Paris, Ramsay, 2003 and Folio, 2005

# Venice

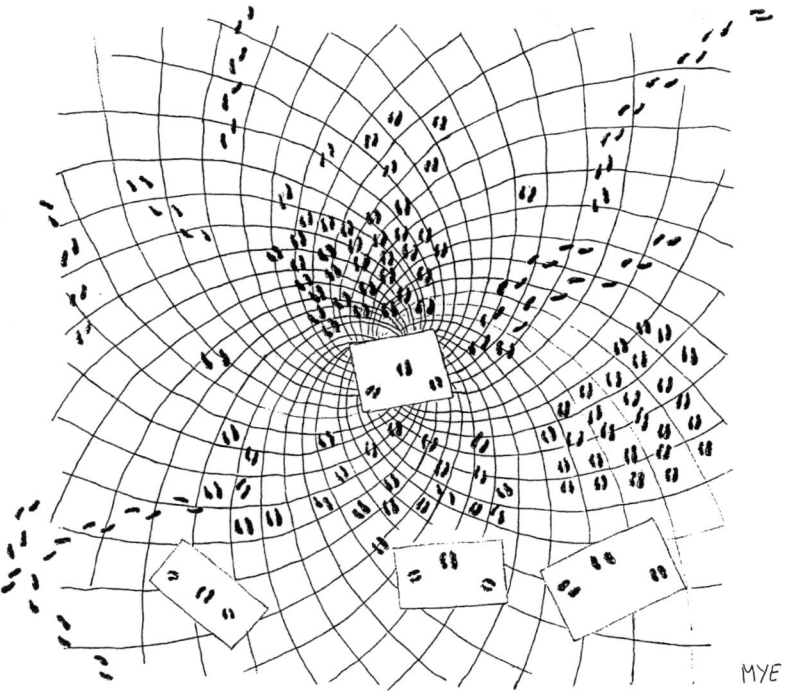

MYE

# The Masks of Venice*

Lina Urban and Guerrino Lovato

*I'll be in front of you and still you won't see me*
*And you'll see me if I stay far away*
*You'll know me by sight and by the colour*
*That always conceals my true appearance*
*And, if you are unsure about my art*
*I will change a hundred shapes away from you*
*And drawing my lights near your eyes*
*I will know how to change again into wild beasts and monsters*

## The Profession of the Mask Makers

In *L'Arte dei mascareri* [1], Lina Urban writes that this poem was inserted into a text that accompanied the only image that has come down to us from the workshop of

* *Translated by Kim Williams*

a mask maker's workshop in Venice: a watercolour by Grevembroch from the second half of the eighteenth century.

In the book's introduction we read:

> If wearing a mask is actually a way of feeling different, a transgression of the rules, for centuries in Venice, in a society in which social barriers existed, an ambiguous face represented the only legal alternative for being equal.

> The mask makers, a subgroup of the painters' guild, were miracle workers who shaped faces for any kind of physiognomy whatsoever, and especially sought after during the six months of the year (Carnival, Sensa, and various public and private festivities) when practically everyone changed their appearances with the complicity of a mask maker.

> In the 1700s the uncontested dominator of costumes was the *tabarro*: the great mantel of black woolen cloth, the tricorn, the three-pointed black hat, and the *bauta*.

> At a moment such as that when the mask had become a genuine consumer good, those working in the profession were few in number: 36 people employed in 12 workshops in 1773.

Also in the introduction, Urban explains how even at that time there were people who worked 'under the table', above all women; now there are probably hundreds of workers, especially in China!

The first *bauta*, the typical Venetian mask, was invented by an anonymous crafts- man, although the use of masks had already been widespread for many years (at least since 1268, since it was in that year that the first laws of the Republic that governed the use of masks were passed). Information about the origins of that mask are found, in particular, in a piece by Pietro Gradenigo in praise of the anonymous mask maker:

> How beautiful the mask invented by the maker of the first *bauta*, because this renders every class and age of person comfortably equal and does not offer the least suggestion as to either gender, even more so because using the black *tabarro* and the white faces today, is completely helpful to economy and liberty in making purchases, in trying one's own ideas, and the only ones to blame are those who change good into evil with that device.

According to the author:

> The profession of the mask makers and the *targheri* (makers of masks and papier-mâché shields) was from the beginning one of the *colonnelli* (specializa- tions in which the painters' guild was divided) established in the guild in 1271. The statute, *mariegola*, from 10 April 1436 to 19 February 1620 is housed in the State Archives in Venice. When the painters split off from the guild, in 1683, constituting themselves as a college, the mask makers were a specialization of decorative painters, along with miniaturists, illuminators, draughtsmen, spe- cialists who worked with painted and gilded leather, gilders, and makers of playing cards.

## The Origins of the Masks

At the 2005 *Matematica e Cultura* conference in Venice, Guerrino Lovato prepared two kinds of masks for the speakers, Bauta and Arlecchino; masks that were made in the 'Mondonovo' workshop a short distance from the conference venue, in campo Santa Margherita. On this occasion Lovato explained the original meaning of the masks:

> In Latin for *maschera* you say *person*, and it is interesting to follow the evolution of this word, which in Italian, to the contrary, designates someone who lives or has lived, that is, a person in flesh and blood; meantime, in France, for example, along with other meanings, *personne* also mean *no one*, as in Latin. Thus a mask was originally a nobody, and so did not represent a particular character, a characteristic human or animal figure, but only the idea of a mask, that is, a face with neither expression nor features, with empty holes for our eyes to see through. The look of the masks that we know today and that are familiar to us derive directly from the painted masks that the Romans used as decorations for their houses and that imitated real masks in terracotta, or more rarely, in marble, called *oscillum*, because they were hung from porticoes and left to swing to keep evil spirits away.

> The expression used in antiquity to indicate this kind of mask could be used indifferently for either comic or tragic masks, which were very different from each other; in any case, they were empty masks, sufficient for the apotropaic aims they were meant to serve.

273

> There is one mask, probably painted in Raphael's milieu, on a portrait cover (a drapery over a painting that could be opened) and accompanied by a motto in Latin *sua cuique persona*, that is, to each his mask. This underlines how the very face of the person portrayed is an illusion, showing in its turn a face that is itself (even outside of the portrait, in reality) a mask, a pose. The Raphaelite mask is not *a* mask, but *the* mask. It doesn't represent someone or something, but purely the game of hiding, of concealment without a costume, without changing ... simply the game of dropping out.

## Papier-Mâché Masks

Today's masks are made in papier-mâché, but their history has undergone various changes over the centuries:

> The history of papier-mâché is tied to the difficulty of conserving it and to its low cost. The major feature of the paper material, besides its economy and its notable plasticity, was its lightness, particularly important for the construction of statues for processions and decorative friezes, which, applied to ceilings in palaces and churches, imitated plaster and marble (the cost of which would have been much higher). Because of the flammability of papier-mâché, which

frequently constituted a fire hazard in churches, its use was prohibited beginning at the end of the 1700s. Our experience, as regards masks, originates in the theatre. At the end of the 1970s in a small workshop in the outskirts of Venice, we again took up the fabrication of professional masks that we used to let dry in the open. Little by little, thanks also to the rediscovery of Venice's Carnival, masks became widespread in the city once again, and a new product was born, which is only apparently the fruit of the history of Venice, but which is really the result of a very recent revival.

Among the many commissions entrusted to Guerrino Lovato was the production of the drawings and models of all the prototypes of the sculpture and bas-reliefs for the ornamentation of the cavea for the reconstruction of the great theatre La Fenice, destroyed by a fire some years ago [2].

## The Technique of Handmade Papier-Mâché

To make a mask it is necessary to begin with a drawing that describes its shape and dimensions. Then can begin the moulding of the clay, a material that is malleable, easy to use, and economical, following the outline of the drawing on the panel.

After having roughed out the model of the mask, defining the volumes of the lineaments, it is smoothed with a little water and polished; next comes the preparation of the imprint mould, that is, the negative upon which the papier-mâché will be worked, to make the filled model, with no eye or nose holes, which will be cut out only after the papier-mâché mask has been taken out of the imprint mould.

At this point, equal parts of quick-setting plaster flakes and water are mixed, and once the right density has been reached (about that of yogurt, to give the idea), the mixture is poured directly onto the clay model, being careful to cover it uniformly.

The plaster solidifies quickly and dries in about an hour, so that it is ready to be carefully separated from the clay. Two days must pass before the papier-mâché work is begun, so that the plaster impression is completely dry. Then work begins by wetting the paper, which has been torn into rectangles, squeezing out the water, and applying it, beginning from the outside and working inwards, letting the borders hang over and being careful to avoid any creases or folds while overlapping the pieces. For the first layer blue recycled rag paper is used; this is more flexible but less strong than paper made of pure cellulose, which will be used for successive layers. Once the first layer is finished, vinyl glue is spread uniformly, and the paper is pressed well in order to make the features of the models clear, then a second layer is added, and a final layer on the edges to reinforce the mask. It is allowed to dry, and when it is completely dry to the touch, it can be detached from the positive imprint; the edges, eye holes and nose holes are then cut with special scissors and blades, and the edges are finished with tissue paper and glue, so that the layers of papers don't come apart. Finally a mixture of equal parts of Bologna plaster and glue is made, which is then spread on the mask as a base for the decoration and which can be smoothed with sand paper to remove any possible imperfections of the papier-mâché.

It should be noted that the plaster used in this phase is completely different from that used to make the imprint mould, which is flaky in texture and very quick-setting. This is Bologna plaster, a product used typically in the fine arts, preferred because it dries very slowly.

At this point the decoration phase begins, with the application of a base of white washable acrylic tempera (two coats are necessary, one denser than the other). Next come the colours, delineating first the eyebrows and then the lips, moles if there are any, a blush of red for the cheeks, and so on, that give the mask its particular character.

The last step is the wax to give it an antique appearance, a practise that derives from the restoration of antique furniture. This is done by applying a compound obtained from a mixture of beeswax and a particular kind of shoe polish with brown or black shades, according to the effect desired; this compound is spread on with a brush and, once dry, polished. The result is like antique wood. After polishing with brushes and dry rags, strings are applied at the level of the eyes and labels, if any, are applied to the inside of the mask.

## Traditional Venetian Masks

### The *Bauta*

The *Bauta*, the Venetian costume par excellence, appeared around 1600.

The term *bauta* does not refer to the face alone, but to the entire costume, which is comprised of the *tabarro*, or mantle, a tricorn, a veil that covers the shoulders, and the *larva*, that is, the actual mask itself. The name *larva* goes back to the Latin; it was used to indicate ghosts and spectral costumes. The *bauta* was a mask that was used without distinction by men and women, and its particular form allowed the wearer to eat and drink while remaining completely incognito.

**Fig. 1.** The *bauta*

## The *Moretta*

The *Moretta* is a mask made of an oval of black velvet that was worn by women of both noble and modest extraction. Its name is derived from *Moro*, which means black in Venetian, and it exalted the white complexions of the women and the Venetian red of their hair.

It is a mask without an opening for the mouth, and was kept on by holding between one's teeth a button placed on the inside.

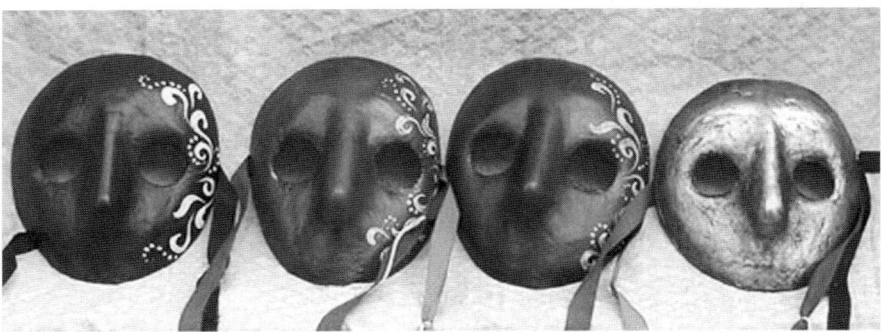

**Fig. 2.** The *Moretta*

## The *Gnaga*

The name *Gnaga* probably derives from *gnau*, the sound a cat makes, and in fact the mask features feline lineaments; it was worn for fun by men with the dresses of courtesans and white bonnets, but also to vent their homosexual tendencies, necessarily repressed at the time by the State Inquisitor.

**Fig. 3.** The *Gnaga*

## The Plague Doctor

This mask was created in the sixteenth century by the French doctor Charles de Lorme. It is not a traditional Carnival mask, but was used as a defence against the terrible plague that struck Venice in 1630. Doctors wore them with black mantles and gloves, and filled the beak of the mask with spices and medicinal essences to neutralise the infectious miasma of the plague.

**Fig. 4.** The Plague Doctor

277

## Pantalone

Pantalone is the best known Venetian mask. It believed to derive from San Pantalon, one of the Venetian saints, to whom a church in the Dorsoduro quarter is dedicated. Pantalone was an old merchant, symbol of the bourgeois and of the old Venetian merchant class. He had a great propensity for business, which sometimes flourished and sometimes drove him to ruin, and a noteworthy nonchalance towards amorous affairs. The mask highlights special somatic features: a hooked nose, prominent eyebrows, and a pointed beard.

**Fig. 5.** Pantalone

## Arlecchino

The harlequin is one of the most popular masks of the Comedy of the Arts, originally from lower Bergamo of the Cinquecento. By nature Arlecchino is a swindler and a busybody, not particularly intelligent, always hungry, and always ready to mooch. The costume is composed of a jacket and pants with coloured squares, a felt hat decorated with a piece of rabbit's or wolf's tail, and a band from which hangs a *batocio*, the spatula used to stir polenta. It is an acrobatic mask, endowed with a rich vocabulary of gestures. The face has features that are demon-like and feline, with a snub nose and a noticeable bump on the forehead, as if to testify to the fact that he always comes out the worse for his adventures.

**Fig. 6.** Arlecchino

## Colombina

Arlecchino's faithful companion, Colombina is a mischievous and astute young maid, also known by the names Arlecchina, Corallina, Ricciolina, Camilla and Lisetta, even following the French mode to become the refined *Marionette* in Carlo Goldoni's *Merry Widow*. Her dress is of coloured squares with an apron and a little

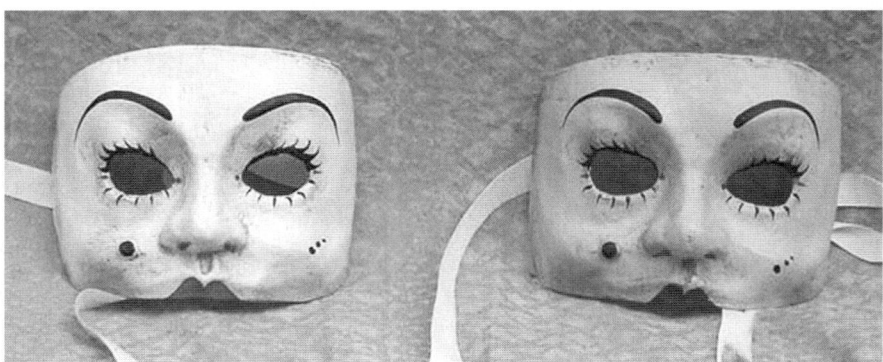

**Fig. 7.** Colombina

white cap. She rarely wears a mask, and if she does, it is only a simple black half-mask that leaves her mouth uncovered. She speaks in various dialects, preferring the Venetian or the Tuscan.

## Bibliography

[1] L. Urban (1989) *L'arte dei mascareri*, Centro Internazionale della grafica, Venice. We thank the author for allowing use of the text.
[2] For all information regarding production of masks by Guerrino Lovato, see *www.mondonovomaschere.it*. Many thanks to Lovato per allowing use of the text and the images.

Mario Merz, *Il volo dei numeri*
(Flight of Numbers), 2000.
Red neon numbers according
to the Fibonacci series. Photo:
Paolo Pellion di Persano, Turin

Colour photograph of the Grand Canal (M. Falcone, pp. 21–32)

A typical filtering test on an image of Lena (M. Falcone, pp. 21–32)

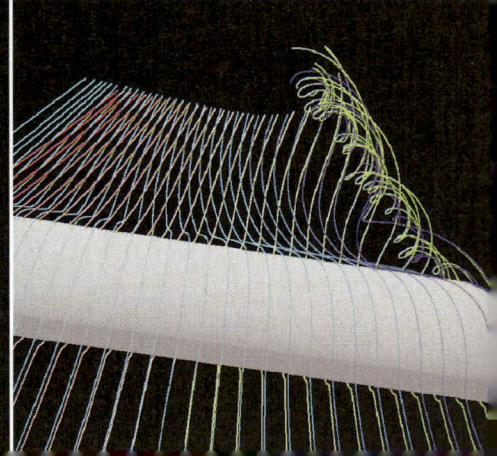

*Above*, Solar Impulse; *right*, vortex at the extremity of the wing of Solar Impulse (A. Quarteroni, pp. 33–45)

*Il pesce quadrato* (The Square Fish) (M. Campana, pp. 121–127)

*Galois nell'ideaspazio* (Galois in ideaspace), © 2005 Paolo Bisi (M. Abate, pp. 147–158)

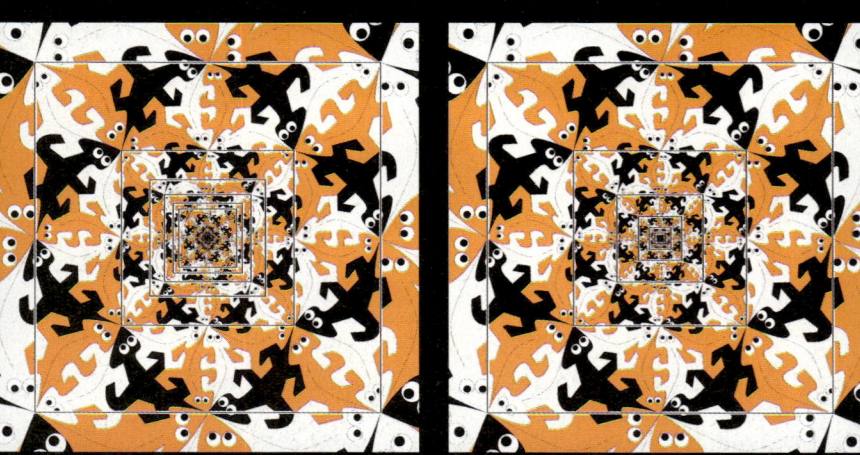

Wassily Kandinsky, *White Cross* (*Weisses Kreuz*), January–June 1922. Oil on canvass, 100.5 × 110.6 cm. Peggy Guggenheim Collection, Venice. (Solomon R. Guggenheim Foundation, NY) (M. Emmer, pp. 173–182)

*At left*, the 'wrong' zoom: scale(t) = A t + 1; *at right*, the 'right' zoom: scale(t) = exp(b t) (G.M. Todesco, pp. 159–170)

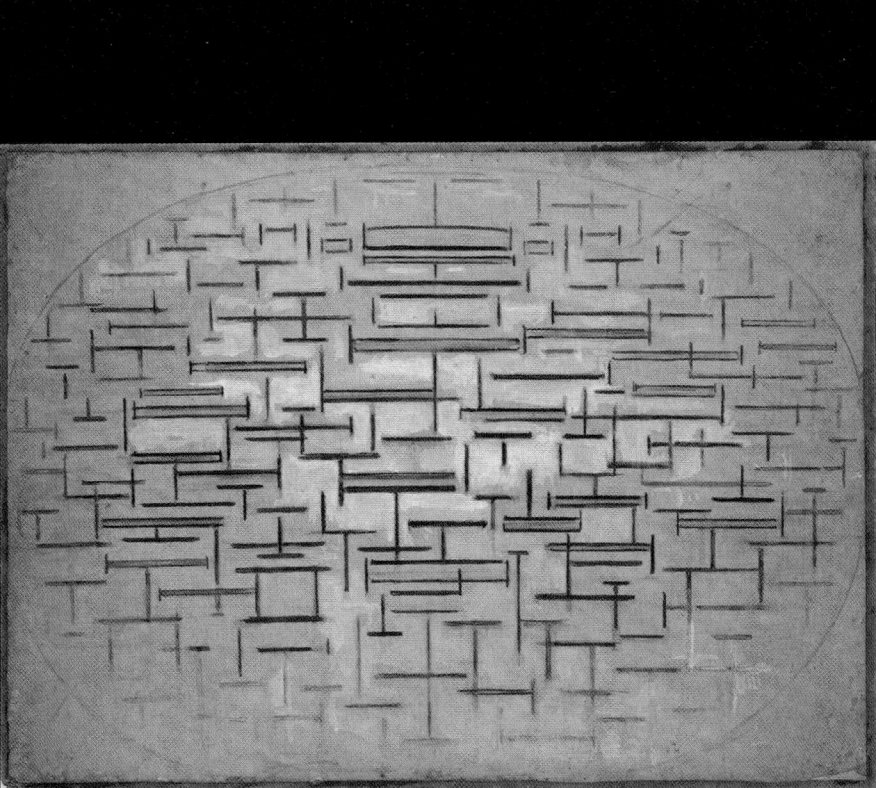

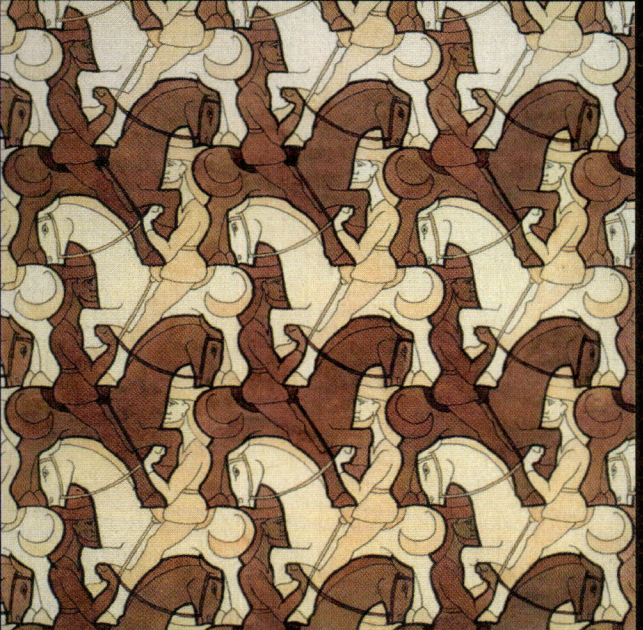

M.C. Escher, Symmetry Drawing E67 (*Horseman*).
 (R. Zilli, pp. 229–238)

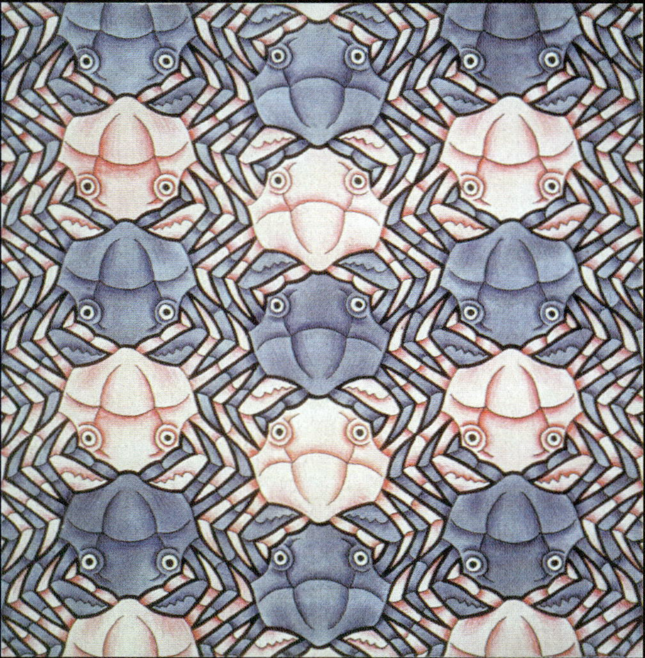

M.C. Escher, Symmetry Drawing E117 (*Crab Canon*).

(R. Zilli, pp. 229–238)

Images of the Mandelbrot set
on the labels of various years
of the wine *Chaos* (A. Terni,
pp. 241–244)

Venetian masks (L. Urban, G. Lovato, pp. 271–279)

# Venice and Marco Polo

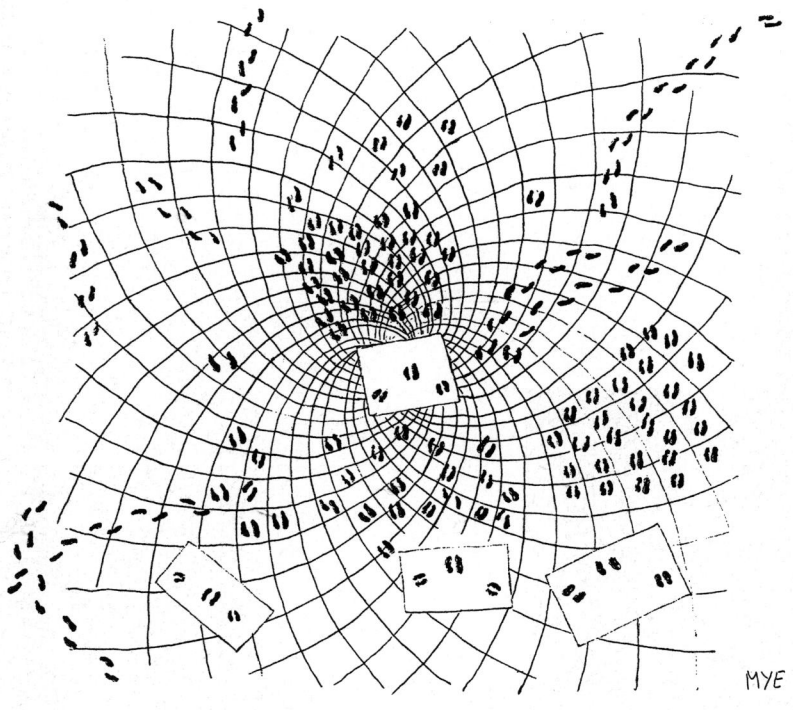

# Introduction*

## Travels with Marco Polo

It is a tradition that participants at the "Matematica e cultura" conferences in Venice are given gifts, gifts that are unique and unrepeatable: from the cartoons of *Lino il Topo* to those of Marco Abate, from small and precious exhibit catalogues from Perilli to Pizzinato, to the small book on *murrine*, and many others. In 2005 the gift was truly exceptional: a copy of the book *In viaggio con Marco Polo* (*Travels with Marco Polo*) [1], with original works of art by forty artists, and three introductory essays. A volume, as always, that had a binding made by hand in the Japanese style, with a cover also made by hand and embellished with spatters of gold: a one-of-a-kind volume, which we hope was appreciated by the participants as it deserved to be. For those who were not at the meeting, we are reprinting here the introduction by Silvano Gosparini and Nicola Sene, as always the curators of all the books of the Centro Internazionale della Grafica, and one of the other papers, that of Michele Emmer, along with some of the plates from the book. We invite you to try to find the original book, because holding it in one's own hands is the only way to do justice to its fascinating beauty.

283

## Bibliography

[1]  Silvano Gosparini, Nicola Sene (eds.) (2005) *In viaggio con Marco Polo*, 40 original works by artists, three introductory essays, Centro Internazionale della Grafica, Venice.

---

*  *Translated by Kim Williams*

# In Venice, Inside Its Grand History*

Silvano Gosparini and Nicola Sene

Without a doubt, for the research work in the field of traditional and experimental graphics uniting the artists of the *Atelier Aperto* – but really, since its inception, the group of artists that make it up – an absolute and primary point of reference has always been the fact of working in Venice. The *Atelier* is situated in the heart of the city (but of course Venice is a city that has many hearts) in the *campo* San Fantin, right next to the great theatre La Fenice, in a typical merchant's house that once hosted George Sand during her Venetian sojourn. This is the confirmation of living inside history.

Artists in general are perceptive and curious, and those of the *Atelier Aperto* especially so, their origins, of the most diverse and mixed kind, extending across Europe to the Americas and Japan, accentuating their peculiarities. In working together they often suggest common themes in order to exercise and compare their own technical knowledge. It was thus that the magical book of the Venetian Marco Polo captured their attention in its byways and traces of the past: the *Corte* of Marco, known as *Il Milione*, the houses of the Polo family, the erratic sculpture, the *cavane* where the ships would depart for the mythical far or near Orient, the fascinating stories … were all sources of inexhaustible inspiration.

The recent collaboration with the passionate cultivators of *Amor del libro* (*Love of books*) has resulted in this edition, which is a kind of narrative through images, a search that, from the original work and the Oriental-style binding, to the historic documents and the poets, has produced a book that is conceived and constructed as in the artistic workshops of centuries ago.

To the whole operation, therefore, goes the credit for having suggesting a rereading of the fantastic book by Marco Polo, in confirmation of our desire to be part of the history of Venice.

* *Translated by Kim Williams*

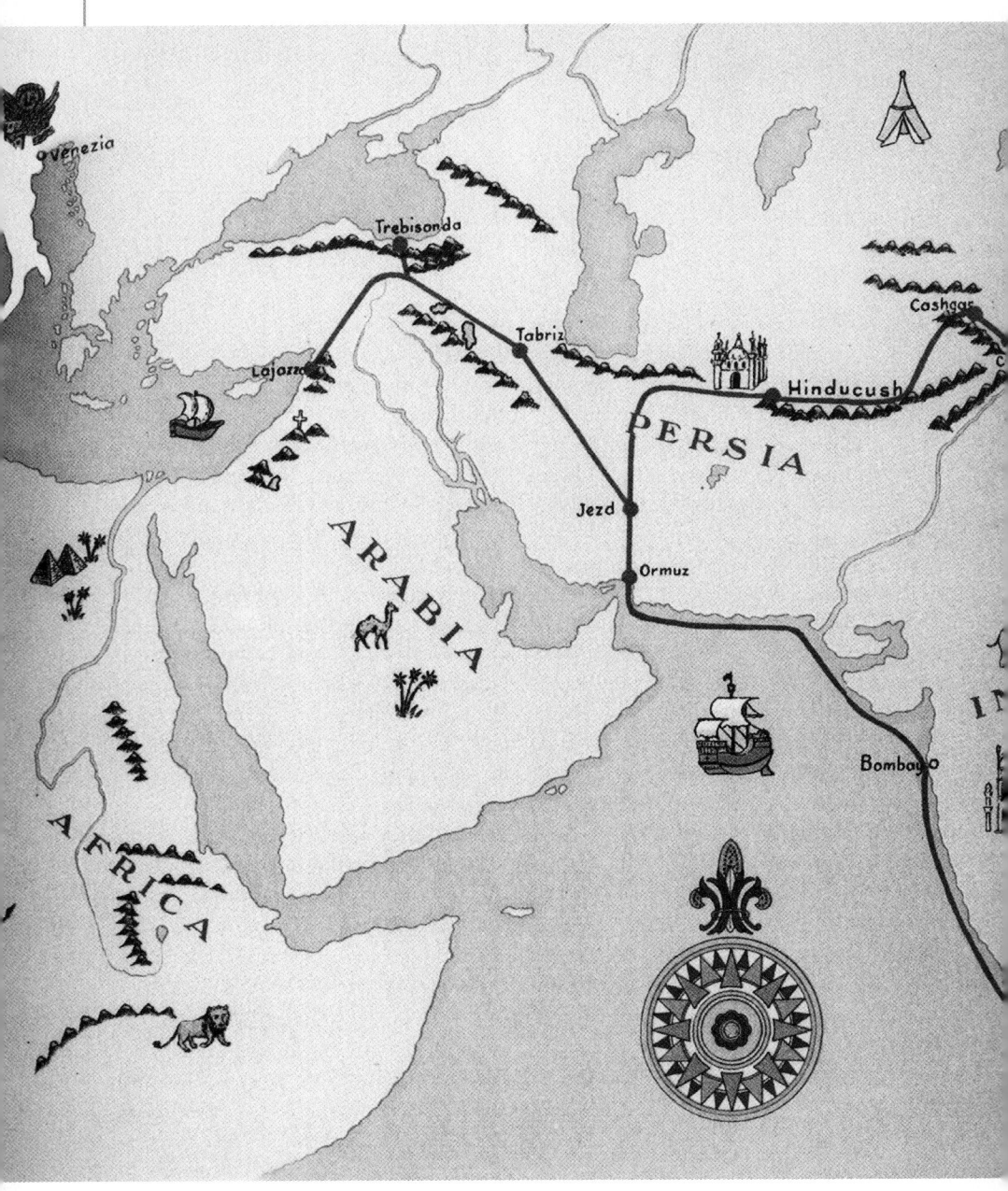

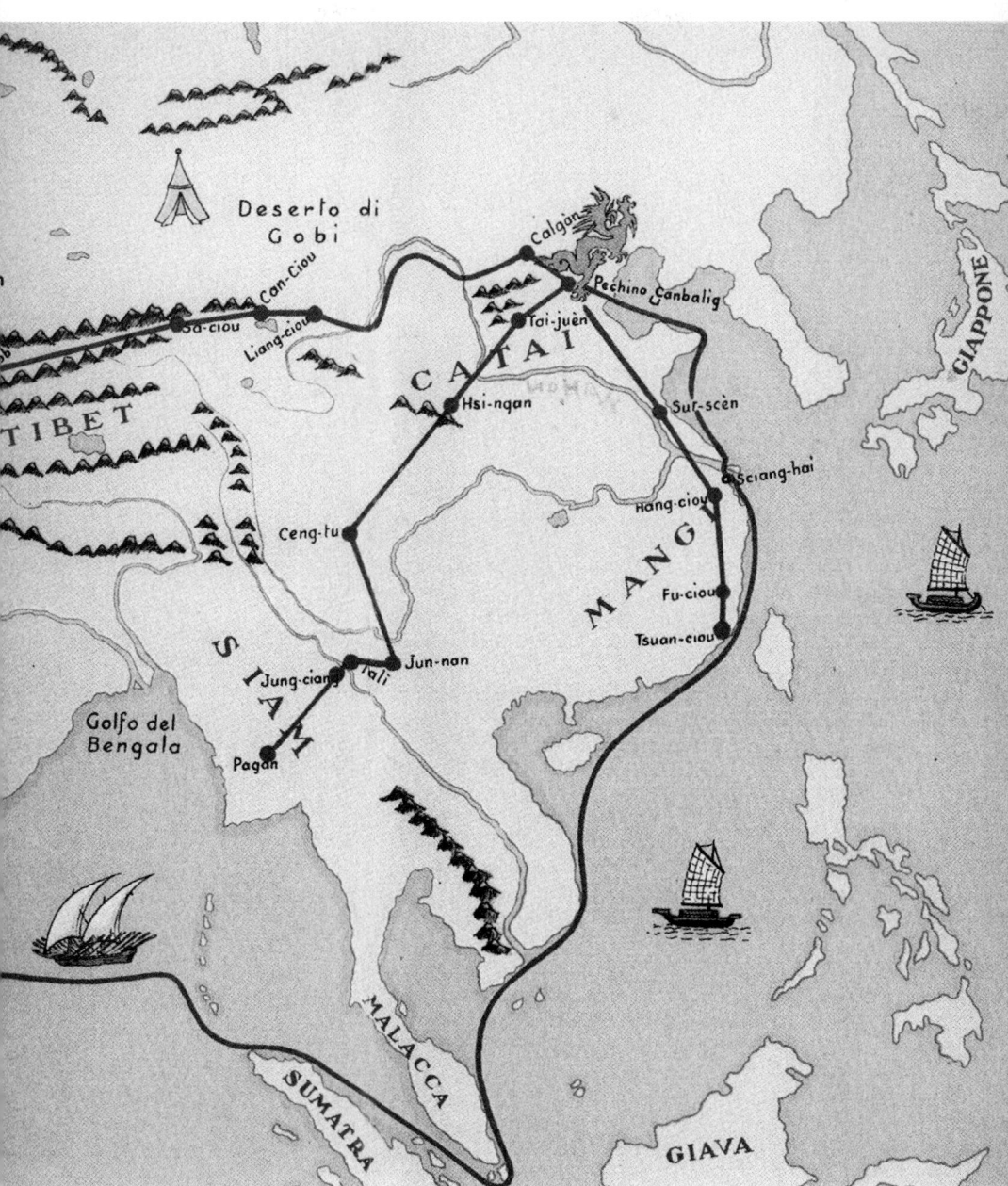

**Fig. 1.** Map of the travels of Marco Polo

# Telling the Wonders of the Discovery of America*

Michele Emmer

> *For ye shall find [in this book] all kinds of wonderful things, and*
> *the divers histories of the Great Hermenia, and of Persia, and of*
> *the Land of the Tartars, and of India, and of many another coun-*
> *try of which our Book doth speak, particularly and in regular suc-*
> *cession.*
>
> The Travels of Marco Polo (The Million), Prologue [1]

From the first to the eighth of August 2001 the famous American cellist of Chinese origin, Yo-Yo Ma, recorded at the Hit Factory Studio a compact disc entitled "Silk Road Journeys", subtitled "When Strangers Meet".

In November 2001, Yo-Yo Ma wrote in the booklet that accompanied the CD:

> To me, this recording is an answer to the question, "What happens when strangers meet?" It is said that when two people meet, within seconds an assessment is made on whether to trust one another. We all know how destructive it is when there is no trust. If there is some trust, an exchange might take place. As this trust develops over time, the exchange may lead to the best of all possibilities – creativity and learning.

> This recording is what happened when 24 strangers, supported by scores of others behind the scenes, met, developed trust, learned from each other, and eventually devised a common language that allowed them to be creative together. We all hope you enjoy it.

Between the time the CD was recorded, in the first days of August 2001, and when Yo-Yo Ma wrote these words passed an infinity of time, the time of the eleventh of September, 2001. The CD recorded by Yo-Yo Ma and the *Silk Road Ensemble* was the result of preparation that began years before. Some musicologists had gathered material in China and central Asia and in July 1999 the *Silk Road* project was launched. Starting in 1999, the *Silk Road Ensemble*, made up of 40 musicians from the United

---

* *Translated by Kim Williams*

States, Europe, Asia and the Middle East had begun to collaborate with each other, to get to know each other, to perform concerts together, up to the realisation of the CD. The work's cover is, of course, a geographic map that goes from Italy to China and Japan and on which are marked all the cities of origin of the players: from Iran to China, from India to Mongolia, using musical instruments that come from the many countries crossed by the mythical Silk Road, the road that seems to have been travelled for the first time from Venice to China by Marco Polo.

Viktor Sklovskij has stated, in a book dedicated to the travelling merchant [2], that he was:

> … fascinated by the destiny of a man who knew how to see, in the Asia of the time, the world of the future. He knew how to describe, without ever lying, concisely and well, Russia. He gave a brief but precise depiction of a whole series of Asian countries, without once expressing his condemnation as a European. He described what he saw, but not what he presumed. I deeply admire his tirelessness, his way of perceiving the world's variety.

From 26 June to 7 July in 2002, some months after the attacks, in the city of Washington D.C. took place "The Silk Road" festival, with the theme "Connecting Cultures, Creating Trust". It was a great venture, with Yo-Yo Ma as inspiration and organized by the Smithsonian Institution Center for Folklife and Culture Heritage. In a Washington park had been constructed a miniature "Silk Road", with artists, craftsman

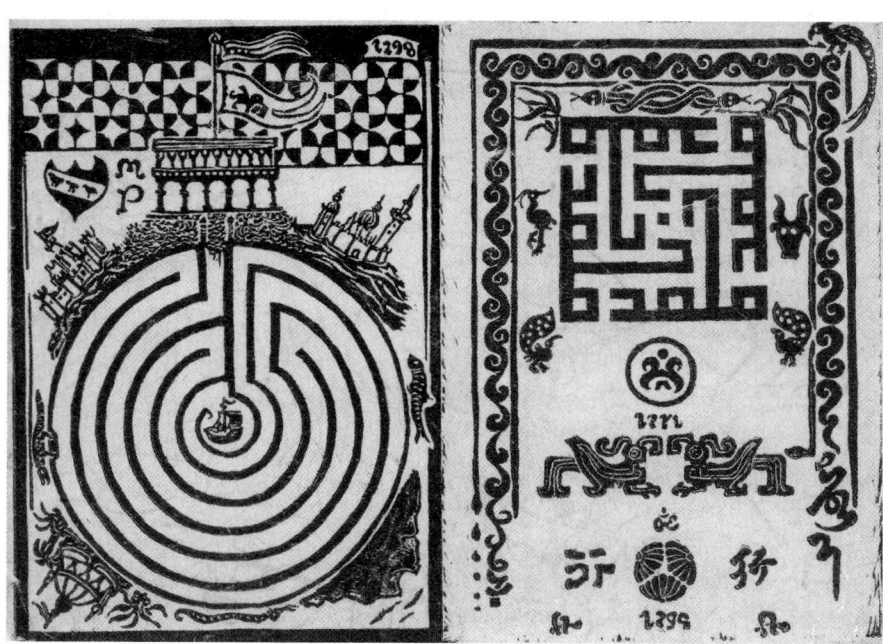

**Fig. 1.** Firenze Poggi, *Viandante nell'intrico lungo del mondo (Wayfarers in the long labyrinth of the world)*

and musicians coming from all of the countries that had been travelled by the merchants, first among them Marco Polo. In order to imagine life in the world at the time of Polo you have to think of the caravan routes: the inhabitants would line up along these or near the ports; thus the people lived in small nests.

In Washington every community was installed in a great tent like the kind used by nomads, so that it was physically possible to meet, in a short stretch of road, the many civilisations, the many languages, the many religions, the many kinds of music, the many foods of a great part of the peoples of Europe and Asia.

In the month of June 2002, in Washington, a city that is of particular importance in this case because it would be where the project "Marco Polo" would be presented for the first time, with artists from various countries of the world gathered around Venice's *Centro Internazionale della Grafica*. Every day for the duration of the "Silk Road Festival" there were performances, concerts, ballets, and talks about the culture, foods, and customs of all the peoples along the mythical Spice Road: from Turkey to Afghanistan, from Iran to Tajikistan, from Turkmenistan to Pakistan, from China to Japan, from Azerbaijan to Armenia, from Caucasia to Syria to Bangladesh. Of course the Italians were present as well, and the Venetians, first of all the musical group "Calicanto", and Guerrino Lovato with his masks [3].

It was a grand encounter of peoples, religions, ethnicities along the Silk Road, with an explicit reference to Marco Polo, not the first to travel the Silk Road but probably the first to cover its entire length. Silk, however, arrived in Rome as early as the first century B.C., where it was a symbol of power and wealth (Julius Caesar would enter Rome triumphantly under silk canopies).

291

**Fig. 2.** Luisa Asteriti, *Partenza da Venice (Departure from Venice)*

Sklovskji writes that the Polo brothers remained for three years in Bukhara; they were merchants and as such they immediately made note of the characteristics of the products of these places:

> [...] The best porcelain in Bukhara comes from China. From China comes silk. From China come objects of gold. Of a woman it is said that she is as beautiful as a Chinese. The goods should be bought there where the prices are low, where they know how to weave, how to bake the white clay [...] The Venetians decided to leave for China [...] The people travelled the great routes not for religion but for commercial affairs. It is precisely this that he spoke of in his book ...

Marco Polo was young and very attracted by the beautiful women:

> These women (of the city of Hangzhou) are very talented and practical in knowing how to flatter and caress with words on the tip of their tongues and made to fit any kind of person, so that the foreigners who experienced them once remain beside themselves, and are so taken by their sweetness and pleasantness, that they can hardly ever forget them ...

and by the inhabitants of the city:

> there is such a degree of good will and neighbourly attachment among both men and women that you would take the people who live in the same street to be all one family. And this familiar intimacy is free from all jealousy or suspicion of the conduct of their women. [...] They also treat the foreigners who visit them for the sake of trade with great cordiality, and entertain them in the most winning manner, affording them every help and advice on their business.

Karl Marx wrote in *Das Kapital*:

> The law that the independent development of commodity capital stands in inverse proportion to the level of development of capitalist production appears particularly clearly in the history of the carrying trade, as conducted by the Venetians, Genoans, Dutch, etc., where the major profit was made not by supplying a specific national product but rather by mediating the exchange of products between commercially – and generally economically – underdeveloped communities and by exploiting both the producing countries.

To be sure, not everyone was like Marco Polo; not everyone is capable of recording and recounting, of running the great risk of not being believed and becoming the laughing stock of Venice upon his return after many years; not many, even in our day, are interested in places, not to mention customs, habits, people, in learning the language, dressing in the manner of the host country, wanting to understand, not being content with just taking.

> Great Princes, Emperors, and Kings, Dukes and Marquises, Counts, Knights, and Burgesses! and People of all degrees who desire to get knowledge of the various races of mankind and of the diversities of the sundry regions of the World, take this Book and cause it to be read to you ...

The book was written for illiterates, as can be seen in the Prologue. These words were written by Rustichello of Pisa, a prison mate of Polo's.

In 2002 photographer Michael Yamashita published the book *Marco Polo: A Photographer's Journey* [4], retracing with his camera the route covered by Polo. An exceptional visual journey that shows how still today the world is of an unimaginable and incredible vastness and variety.

In that same year the idea of the "Marco Polo" project was born among the artists who gravitated around the Centro Internazionale della Grafica in Venice, a place (because places are important, and will never be substituted by virtual sites) along one of the silk routes that begins and ends in Venice. A place, but one made of people, as Marco Polo, who wrote of men and women, knew so well, fascinated by their appearances, their clothes, their customs, and their languages. A collective of artists who orbited around one of the Venetian places that is a magnet for the graphic arts; a *salon*, we might say, one that has a long experience of community activities, one that for years has facilitated the exchange between different cultures (the unforgettable project *"Gioco del Pesse"*, a game, exhibit and day in the city, following a kind of "silk road" through Venice [5].

And the Marco Polo project, which, for reasons that were mysterious but inevitable, could only open to the public in Washington, a city that should be added *ad honorem* to the places of the Silk Road, once again brings together many artists in a place that might have appeared to be virtual (not everyone had the possibility of

**Fig. 3.** Walterina Zanellati, *In viaggio con Marco Polo (Travelling with Marco Polo)*

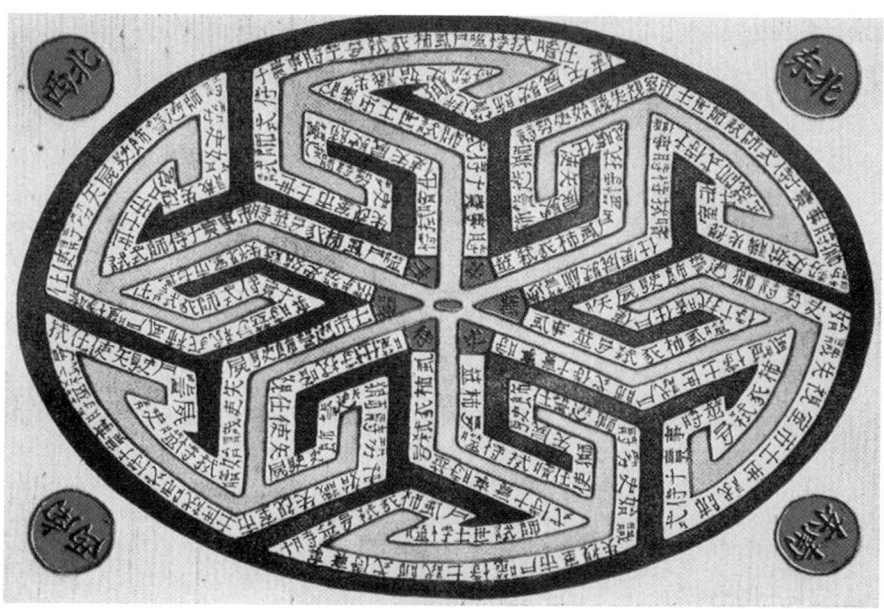

**Fig. 4.** Tiziana Talamini, *Mondo (World)*

actually travelling the Silk Road like Yo-Yo Ma and Michael Yamashita did) but was not virtual at all in fact, since, as Marco Polo taught us, from that imaginary – but much more real – journey along Polo's route the artists brought back images which in their turn create still another journey. And that this should happen in the Americas was a sign of destiny. When Marco Polo's book began to be accepted, not as a book of fables but as the recollections of a visitor to those faraway lands (and that acceptance would take decades and decades), maps were drawn and lines of communication opened. It was precisely by reading that book that Cristoforo Colombo became convinced that he could reach faraway Asia by sailing due west, without having to circumnavigate Africa. So, turning towards the islands of the Caribbean, he believed that he had arrived in Asia and wrote letters to the Great Khan of China. As Sklovskij writes:

> Thus the truthful book of Marco Polo earned the fame of lying to travellers, and an error, inspired by this book, led another traveller to discover America, which he mistook for Asia.

The Silk Road has become, thanks to "The Million", a sign (a graphic sign, you might say) of how humanity longs to know itself, spend time with itself, love itself. A utopia, to be sure, but which in Venice, a city where everything begins and everything ends, only to begin again (aren't the tides a testimony to this?) it seems less elusive.

# Bibliography

[1] Marco Polo and Rustichello di Pisa (2004) *The Travels of Marco Polo*. Trans. Henry Yule (1903), notes and comments by Henri Cordier. Project Guttenberg, EBook #10636, http://www.gutenberg.org/files/10636/10636-8.txt.

[2] V. Sklovskij (1972) *Marco Polo*, Il Saggiatore.

[3] L. Urban, G. Lovato (2006) *The Masks of Venice*, in the present volume.

[4] M. Yamashita (2002) *Marco Polo: un fotografo sulle trace del passato*, Edizioni White Star (Eng. Trans. *Marco Polo: A Photographer's Journey* (2004), White Star).

[5] S. Gosparini, ed. (2002) *Il gioco del Pesse*, 46 artists, three introductory texts, Centro Internazionale della Grafica, Venice.

# Authors

Marco Abate — Department of Mathematics, University of Pisa

Giovanni Maria Accame — Accademia di Brera, Milan

Carlo Boldrighini — Department of Mathematics, University of Rome "La Sapienza"

Elisa Cargnel — Akademia Olympia, Venice

Marcus Du Sautoy — Mathematical Insitute, Oxford, UK

Michele Emmer — Department of Mathematics, University of Rome "La Sapienza"

Maurizio Falcone — Department of Mathematics, University of Rome "La Sapienza"

Loe Feijs — Department of Industrial Design, Eindhoven Technical University, The Netherlands

Davide Ferrario — Film director, Torino, Italy

Manuele Gandini — Art Critic, Italy

Silvano Gosparini — Centro Internazionale della Grafica, Venice

Robert Kanigel — MIT, Boston, USA

Marco Li Calzi — Department of Applied Mathematics, University "Ca' Foscari", Venice

Guerrino Lovato — Mask maker, Venice

Massimo Marchiori — University "Ca' Foscari", Venice

Maria Rosa Menzio — Playwright, Torino

| | |
|---|---|
| **Maria Cristina Molinari** | *Department of Economic Sciences, University "Ca' Foscari", Venice* |
| **Giovanni Naldi** | *Department of Mathematics "F. Enriques", University of Milan* |
| **Nicola Parolini** | *École Polytechnique Fédérale de Lausanne, Switzerland* |
| **Marc Petit** | *Author, Paris* |
| **Christophe Prud'homme** | *École Polytechnique Fédérale de Lausanne, Switzerland* |
| **Alfio Quarteroni** | *École Polytechnique Fédérale de Lausanne, Switzerland MOX, Department of Mathematics, Politecnico di Milano, Italy* |
| **Gianluigi Rozza** | *École Polytechnique Fédérale de Lausanne, Switzerland* |
| **Antonello Sciacchitano** | *Psychiatrist and psychoanalyst, Milan* |
| **Nicola Sene** | *Centro Internazionale della Grafica, Venice* |
| **Victor Simonetti** | *Architect, visual operator, Pieve Ligure* |
| **Fabio Spizzichino** | *Department of Mathematics, University of Rome "La Sapienza"* |
| **Antonio Terni** | *Oenologist, viticulturist, Azienda agricola "Le terrazze", Numana, Ancona, Italy* |
| **Gianmarco Todesco** | *Digital Video s.r.l., Rome* |
| **Lina Urban** | *Centro Internazionale della Grafica, Venice* |
| **Adolfo Zilli** | *Film director, Akademia Olympia, Venice* |

# Mathematics and Culture Collection

Emmer, M. (Ed.)
Mathematics and Culture I
ISBN 978-3-540-01770-7, VIII, 352 pages, 2003

Emmer, M. (Ed.)
Mathematics and Culture II. Visual Perfection: Mathematics and Creativity
ISBN 978-3-540-21368-0, X, 203 pages, 2005

Emmer, M. (Ed.)
Mathematics and Culture III
with CD-ROM, approx 260 pages, ISBN 978-3-540-34259-5, due 2009

Emmer, M. (Ed.)
Mathematics and Culture IV
ISBN 978-3-540-34254-0, VIII, 253 pages, 2007

Emmer, M. (Ed.)
Mathematics and Culture V
ISBN 978-3-540-34277-9, X, 269 pages, 2007

Emmer, M. (Ed.)
Mathematics and Culture VI
ISBN 978-3-540-87568-0, X, 299 pages, 2009

Additional information on these volumes can be found at http://www.springer.com

Printing: Krips bv, Meppel, The Netherlands
Binding: Stürtz, Würzburg, Germany